How to Do Things with Sensors

Forerunners: Ideas First

Short books of thought-in-process scholarship, where intense analysis, questioning, and speculation take the lead

FROM THE UNIVERSITY OF MINNESOTA PRESS

How to Do Things with Sensors

Jennifer Gabrys

University of Minnesota Press

MINNEAPOLIS

LONDON

The research leading to these results has received funding from the European Research Council under the European Union's Seventh Framework Programme (FP/2007-2013)/ERC Grant Agreement no. 313347, "Citizen Sensing and Environmental Practice: Assessing Participatory Engagements with Environments through Sensor Technologies."

Published by the University of Minnesota Press, 2019
111 Third Avenue South, Suite 290
Minneapolis, MN 55401-2520
http://www.upress.umn.edu

Available as a Manifold edition at manifold.umn.edu

The University of Minnesota is an equal-opportunity educator and employer.

Contents

Introduction: How to Do Things with Sensors

THE WORLD OF SENSORS is one of amplified connections. Sensors are meant to join up and speed up, while also facilitating and enabling. Whether these functions pertain to adjusting lighting levels or advancing political engagement, a quickening of activity is expected to unfold through sensors. Sensors are embedded in urban infrastructure and surveillance systems, and they are also packaged as makerly kits and citizen-sensing projects. But to get here, you need to follow the instructions. Working with sensors typically involves the uptake of the how-to guide. Citizen sensing requires settling into the instructional mode and the imperative mood. A sensing citizen is a handy subject, an action-oriented and technically equipped actor able to tinker toward new configurations. Behold, the Instrumental Citizen.

Sensors, and the growing use of these devices for citizen-sensing projects, are in some ways part of a wider movement toward the how-to and the do-it-yourself. Handbooks and user's guides cover topics both technical and philosophical. From YouTube videos providing instruction on how to troubleshoot the use of microcontrollers to handbooks for using and "abusing" the Internet of Things (IoT)[1] to how-to forums and Instructables, the genres and formats of the digital continue to expand and develop into well-used vehicles for technical instruction. How-to guides and instructions are integral to computation. Code is an unfurling of instructions. Algorithms set in place procedures for

1. See Nitesh Dhanjani, *Abusing the Internet of Things: Blackouts, Freakouts, and Stakeouts* (Beijing: O'Reilly, 2015).

computational processes. The instructional has intensified its residency in machines.

As the how-to proliferates, and instructions unfold through every aspect of the computational, it would seem worthwhile to ask why the how-to has become one of the prevailing genres of the digital. Why the guide, why now, why in this format? Alongside wondering about the how-to format, this text examines how the instructional approach to sensors contributes to particular ways of inhabiting these devices. *How to Do Things with Sensors* is as much a question as a set of instructions that asks how things are made do-able with and through sensors. As a guide, it considers how worlds are made sense-able and actionable through the instructional mode of citizen-sensing projects.

Although the how-to guide is now prevalent within digital spaces, it is certainly not unique to the realm of the digital. Handbooks advise how to live forever, how to live on Mars, how to meet aliens, how to conquer the internet, how to make a million, how to build a rocket, how to clone a sheep, how to split the atom, and how to save the planet. And this list is just a cursory scan. How-to guides run into the tens of thousands. The how-to guide not only outlines procedures for attaining the grandiose or epic but also informs on the banal and the necessary. There are how-to guides for foraging for food after a disaster and for learning handicrafts. How-to guides also provide instructions for organizing political campaigns, for undertaking direct action, and for following step-by-step programs toward greater democratization.[2]

2. A vast range of texts provide resources for social organizing and participation, from the straightforward handbook of Matthew Bolton, *How to Resist: Turn Protest to Power* (London: Bloomsbury, 2017), to detailed considerations of how participation is not evenly available—and how to change the conditions of political engagement through guides that take the form of the syllabus, as in Matt Chrisler, Jaskiran Dhillon, and Audra Simpson, "The Standing Rock Syllabus Project," October 21, 2016, http://www.publicseminar.org/2016/10/nodapl-syllabus-project/#.WEiq1KIrIxG. For a related critique of discourses of participation, see Jaskiran Dhillon's

One particular subgenre of the how-to guide that is of particular interest for the purposes of this study is what might be considered a more radical or countervailing approach to instructions. These how-to guides span from instructions for surviving in a world with which one is at odds to how-to instructions for working and re-working digital technology toward engagements beyond the usual privileged actors and how-to strategies for making worlds that are more livable.[3] These how-to guides are far from simplistic in their assessment of the problem at hand or in the remedying approaches that might be attempted. Instead, they often undertake a necessary and powerful diagnosis of inequality, injustice, and distress and of how to combat or circumvent these conditions. In this way, the how-to guide is not merely instructional for assembling gadgets; it is also a life guide, suggesting how best to carry on. Still other how-to guides are written as counterinstructionals. Joanna Russ's *How to Suppress Women's Writing* uses the how-to guide as a way to demonstrate the multiple strategies by which women's writing is dismissed, derided, appropriated, and erased. Her point is not to facilitate or encourage these practices but rather to expose the habitual, recurring, pernicious, if often implicit ways in which some voices are excluded so as to amplify the writings of the privileged. How-to guides express political commitments. They are not merely a universal set of skills for anyone to follow, even for a seemingly accessible technology such as citizen-sensing devices. They can also serve as normative devices, reproducing unjust political structures and relations. Or they can provide resources for breaking with and addressing inequities.

How-to guides often are organized as or accompanied by tool kits that serve as the essential components for configuring and ma-

---

Prairie Rising: Indigenous Youth, Decolonization, and the Politics of Intervention (Toronto: University of Toronto Press, 2017).

3. Sara Ahmed outlines strategies for surviving in a world with which one is at odds, as well as ways to build other worlds, in *Living a Feminist Life* (Durham, N.C.: Duke University Press, 2017).

terializing instructions. Tool kits are the thing to be constructed, gathered, and mobilized; they are the instrument to achieve particular effects. Astonishing capabilities are often channeled through tool kits, from suitcases and bug-out bags that can save the world (or save oneself at the end of the world)[4] to ad hoc solar power setups and moonshine installations.[5] At least since the 1960s, how-to guides have been bound up with a seemingly countercultural ethos, where access to tools might rework living environments.[6] Some of these kits and instructions yield the unprogrammatic, the unexpected, the incommensurate, and the incomputable.[7] Other DIY projects present "tactical" strategies for intervening within the current operating systems of technoscience, to realize more socially just or equitable relations.[8]

4. For just a few examples of these projects, see El Recetario (The cookbook), http://el-recetario.net/; Makea Tu Vida, http://www.makeatuvida.net/; and "What You'll Need to Escape New York," *New York Times,* January 25, 2013, https://archive.nytimes.com/www.nytimes.com/interactive/2013/01/27/nyregion/preppers-bug-out-bag.html.

5. See Charles Jencks and Nathan Silver, *Adhocism: The Case for Improvisation* (1972; repr., Cambridge, Mass.: MIT Press, 2013); Michel Daniek, *Do It Yourself 12 Volt Solar Power* (East Meon, U.K.: Permanent Publications, 2007); Eliot Wigginton and His Students, *The Foxfire Handbook* (New York: Anchor Books, 1972).

6. The *Whole Earth Catalog* might be the usual reference here, as discussed in Andrew G. Kirk, *Counterculture Green: The "Whole Earth Catalog" and American Environmentalism* (Lawrence: University Press of Kansas, 2007), and Fred Turner, *From Counterculture to Cyberculture: Stewart Brand, the Whole Earth Network, and the Rise of Digital Utopianism* (Chicago: University of Chicago Press, 2006). During the early 1970s in Italy, a parallel and radical approach to technology developed with the Global Tools project, documented in the collection edited by Valerio Borgonuovo and Silvia Franceschini, *Global Tools: 1973–1975* (İstanbul: SALT/Garanti Kültür, 2015), http://saltonline.org/media/files/globaltools_scrd.pdf.

7. For an example of the serendipity that can emerge through tool kits in the form of Fluxus projects, see Hannah Higgins, *Fluxus Experience* (Berkeley: University of California Press, 2002).

8. Beatriz da Costa and Kavita Philip, eds., *Tactical Biopolitics: Art, Activism, and Technoscience* (Cambridge, Mass.: MIT Press, 2008); Alondra

Within this realm of the instructional, it seems the manifesto has given way to the tool kit. The manifesto form of often dogmatic proclamations has yielded to a more open-ended organization of practices. Radical tool kits in particular gather instruments and resources that serve as practical philosophies and modes of organizing in difficult circumstances. There are speculative "feel-tank kits" that are meant to retool political engagements and imaginaries; "killjoy survival kits" for living a feminist life and surviving ongoing adversity; and radical pedagogy kits for sharing skills, lore, and community-organizing techniques.[9] At the same time and in a diverging way, policy recommendations now often circulate in the form of tool kits, a format that governments and NGOs use to advocate for the accessibility of the policy-making process.[10] Such an approach runs the risk of shoring up power structures, while presenting governance as a more transparent and DIY process. In this way, tool kits can be shape-shifting genres that variously provide guidance for rerouting, or reinforcing, sociopolitical practices.

While the focus here is on the how-to in relation to digital technologies, especially environmental sensors and citizen-sensing projects, an extensive assortment of instructionals and tool kits then informs this study. A history of how-to and instructional guides would be an interesting project indeed, although this is not within the scope of this research. Instead, I examine the instructions and instructional approach that inform and develop through the configuration and use of DIY sensing technologies.

Nelson, *Body and Soul: The Black Panther Party and the Fight against Medical Discrimination* (Minneapolis: University of Minnesota Press, 2011).

9. The Feel Kit is a speculative project from Feel Tank Chicago, described at http://feeltankchicago.net/. Sara Ahmed describes the Killjoy Survival Kit in *Living a Feminist Life*, 235–49. The art and research program How to Work Together is an example of a multiplatform collaborative project investigating alternative modes of community organization and collaboration, available at http://howtoworktogether.org/.

10. For instance, see the U.K. government's Open Policy Making Toolkit, https://www.gov.uk/guidance/open-policy-making-toolkit.

In working through the instructional approach, I consider in what ways how-to guides not only enable technical engagements with citizen-sensing devices but also provide distinct material formats and political practices for addressing environmental problems. Tool kits offer ways not just to make sensors but also to construct social–political worlds.

The procedural approach could be seen to promise an outcome of sorts. Yet what happens when instructions are imperfectly followed or ignored? And what occurs if the promised effects do not unfold as expected? The how-to guide could be encountered at once as a set of helpful instructions and as a potentially overly programmatic mode of engagement. The how-to as easily moves from the how-to-do to the do-it-this-way-or-else. The how-to could be a reversioning of cybernetic command and control: follow these steps to a certain outcome, not through an overarching program of control, but through your own pursuit of edifying instructions. Here technology could close in on itself, and instrumentality could lead to the bad sort of functionality that philosophers engaged with technology have warned against. Technologies in this register are seen to fulfill mere functions, and as Simondon has suggested, this master–slave condition can overlook or suppress the relations that unfold across environments, subjects, and zones of energetic and cultural transfer.[11]

The how-to could then be seen to be expressive of an *imperative mood*. The procedure and the following of instructions are bundled up with a pedagogy infused with soft commands and imperative verbs. The digital engenders a directional approach. The

11. Gilbert Simondon, *On the Mode of Existence of Technical Objects,* trans. Cécile Malaspina and John Rogove (1958; repr., Minneapolis, Minn.: Univocal, 2017). Much more could be written about Simondon's specific discussion of instrumentality and the master–slave dynamic within technology, a topic that has also been discussed at length by scholars of race and technology, including Louis Chude-Sokei, *The Sound of Culture: Diaspora and Black Technopoetics* (Middletown, Conn.: Wesleyan University Press, 2015) (thanks to Louis Henderson for this reference).

how-to makerverse assembles as a hub of instruction. Image tiles organize into task sessions; search bars open up into multidimensional cosmologies of the how-to; the ten-minute step-by-step guide promises incremental if expedient accomplishment; and the soothing virtual confab of video narrators begins with the ever familiar entry point "Hi, guys, I'm here to show you how to . . ." The digital and instrumental citizen who enters the process of the how-to is not just learning how to make and use technology but is also entering into modes of procedure, instruction, implementation, and instrumentality that guide digital participation.

This text engages with the how-to and the imperative mood as key aspects of how citizen-sensing technologies and practices are made and organized. *How to Do Things with Sensors* is, on one hand, written as a response to the many requests I have received to write a how-to guide for undertaking citizen sensing. It provides a handbook—of sorts—with tips and pointers for carrying out citizen sensing, while recounting experiences of initially testing sensors for environmental monitoring and then deploying these in the field. On the other hand, I also reflect on the how-to as a genre and approach and ask, Why is this mode seemingly most effective for engaging with digital technologies? What are the opportunities of the how-to approach, what are its fizzy promises, and what are its unseen pitfalls? Rather than describing the phenomenon of the how-to through a distanced commentary, however, I work through instructions and approaches to building kit, to monitoring environments, to analyzing data, to communicating evidence, and to attempting to realize political change through a collaborative project that I lead, Citizen Sense, which has researched citizen-sensing technologies and practices since 2013. I consider how protocols and tool kits, practical manifestos and political programs, inform and materialize as citizen-sensing projects and practices. By attending to the how-to guide as a particular orientation to technology, I suggest that engaging with tool kits and guides is also a way of working through, reworking, and transforming the possibilities of technical, political, and environmental practice. In this way, I ask how an engagement with tool kits can become a way to *retool* approaches to instruments, instrumentality, and digital worlds in the making.

The practices involved in undertaking citizen-sensing projects require not just putting together assorted electronics but also attending to complex configurations of technology, politics, environments, and modes of citizenship. This examination of the how-to and the tool kit does not reify making and craft but rather offers a theory of practice and action oriented toward change. This text attends to that which is often left out of many how-to guides. By engaging with what might be seen to be on the margins of technical interest, I hope to rework how citizen-sensing technologies are usually encountered, less as instruments able to implement certain ends, and more as openings into rethinking socioenvironmental potential and technopolitical relations. Can instruments generate instrumentality, not of the positivist sort, but more of the pragmatist type, where as John Dewey has suggested instrumentality necessarily requires an experimental and contingent set of engagements?[12] Instruments as ideas require what William James has called the "open air" to unfold as lived experiences and processes. Indeed, as Cornel West has pointed out, pragmatism concerns the contingencies of subjects, collectives, and the world as well as theories and knowledge, which in their more pliable formation could be able to respond to social crises and democratic struggles.[13]

By reading the pragmatists sideways through engagements with feminist technoscience and indigenous and critical race theory, and by working with instruments in practice, I suggest that it might be possible to reengage with instrumentalism beyond its usual extractive and expedient registers to consider expanded relations of effect and effectiveness. I situate this engagement within the open

12. John Dewey, "The Development of American Pragmatism," in *The Later Works, 1925–1953,* vol. 2, *1925–1927,* 3–21 (Carbondale: Southern Illinois University Press, 1984).

13. Cornel West, *The American Evasion of Philosophy: A Genealogy of Pragmatism* (Madison: University of Wisconsin Press, 1989).

air of inquiry to express the sociopolitical constitution of instruments as much as to relay how devices are situated in and make worlds. The how-to is a proposition for open technology, which, as Gilbert Simondon suggests, can be a way to engage with machines beyond fixed outcomes as well as an opening into alternative configurations of humans, nonhumans, and relations.[14] "How-to" here becomes an invitation to make, organize, orchestrate, conjure, and sustain people, technology, and worlds toward openings rather than prescribed ends.[15] While ends might inform the starting points for particular technopolitical practices, they inevitably change along the way. An instrumental proposition becomes a site of transformation. This guide proposes how to retool the how-to approach, not to proceed toward certain outcomes, but rather to work for open engagements. I call this approach *open-air instrumentalisms*. Retooling is a practice in open-air instrumentalism.

14. Simondon, *On the Mode of Existence of Technical Objects*.

15. This discussion picks up where my related earlier text, *Program Earth*, left off in thinking about propositions for open technology (which are somewhat different expressions of "openness" than those that call for open hardware, software, and data—as this openness requires an attention to the milieus that inform technology as an expanded field of relations). See Jennifer Gabrys, *Program Earth: Environmental Sensors and the Making of a Computational Planet* (Minneapolis: University of Minnesota Press, 2016).

How to Construct Tool Kits

MANY DIGITAL TECHNOLOGIES occupy a curiously contrary position. On one hand, they seem to offer engagement and empowerment for individual users. On the other hand, they are often highly controlled and monopolized technologies that bind users into particular practices and relations. Such an observation has been made through numerous analyses of an array of digital gadgets, data, and platforms. Digital technologies organize devious relations to a seemingly empowered if minutely surveilled user, from social media that profiles users while enabling views to be broadcast to apps and wearables that track sleep and fitness for the promise of greater well-being and productivity. Digital technologies manage, inform, and otherwise mediate everyday activities, and these functions are overseen by a limited number of organizations with often questionable agendas.[1]

To unravel the typically rigid contours of digital devices, many technology advocates have begun to assemble microcontrollers and sensors and code to develop a more informed engagement with these machines. Here the tool kit becomes a way to fashion a more deliberate encounter with digital devices. By making electronics from the ground up, one is meant to be able to understand the decisions that are made in setting up technical configurations one way and not another. But even the fashioning of a tool kit al-

1. For instance, see Ruha Benjamin, ed., *Captivating Technology: Race, Carceral Technoscience, and Liberatory Imagination in Everyday Life* (Durham, N.C.: Duke University Press, 2019). Benjamin's collection captures the "carceral techniques" that have been implemented in policing, prisons, surveillance, and profiling and yet are also critically engaged with to forge potential sites of retooling and liberation.

ready contains a set of built-in assumptions and orientations to-ward technopolitical action. It is this unboxing and remaking into tool kits for assembly that I examine more closely here. Multiple tool kits address how to develop observant digital practices, including how-to guides for erasing your internet profiles, how to become anonymous online, how-to instructions for undergoing a data detox, and how-to guides for building a DIY bulk surveillance system.[2] In this sense, tool kits not only provide instructions and materials but also indicate how to live with, through, and against these technologies. Tool kits provide instructions not just for assembly and use but also for attending to the social and political ramifications of digital devices.

Citizen-sensing technologies present a similarly complex set of instructions, practices, relations, and politics. Many of these technologies monitor environmental variables, such as air or noise pollution. Air quality monitoring tool kits can be found in multiple forms, which do not even necessarily include digital devices. Indeed, there are long-standing practices of working with diffusion tubes to monitor nitrogen dioxide. Small plastic tubes can be affixed to lampposts and street signs across an urban area and then sent to a laboratory for analysis. These low-cost analog devices provide a monthly average of nitrogen dioxide levels and

2. For examples of these guides, see Andrew Tarantola, "How to Erase Yourself from the Internet," *Gizmodo,* November 2, 2013, https://gizmodo.com/how-to-erase-yourself-from-the-internet-1456270634; LA Crypto Crew, "How to Become Anonymous Online," *Hyperallergic* (blog), December 2, 2016, https://hyperallergic.com/342262/a-guide-to-becoming-anonymous-online; Tactical Tech, "Data Detox Kit," produced for the Glass Room, London, November 2017, https://datadetox.myshadow.org/en/detox; Kim Zetter, "How to Make Your Own NSA Bulk Surveillance System," *Wired,* January 27, 2016, https://www.wired.com/2016/01/how-to-make-your-own-nsa-bulk-surveillance-system. This is a short list that could be significantly expanded. For instance, see also *Bellingcat*'s multiple how-to guides, including Nathan Ruser, "How to Scrape Interactive Geospatial Data," September 5, 2018, https://www.bellingcat.com/resources/how-tos/2018/09/05/scrape-interactive-geospatial-data.

can give an indication of pollution levels in an area. These monitoring practices often also include instructions for how to undertake a campaign for improving air quality, together with technical instructions for installing diffusion tubes and analyzing data. The tool kits that assemble along with these devices can be oriented to community organizing and local activism, urban design and traffic interventions, collective mappings and town-hall meetings, and problems of digital functionality. These aspects of tool kits are no less important, yet they do have a tendency to recede from view when the focus is on learning the technical aspects of digital monitoring technologies.

Citizen sensing is another set of such technologies and practices for monitoring the air that are often bundled into tool kit form. These digital technologies are used to monitor and measure environmental problems and to generate data that could be actionable for policy and regulation. The rise of citizen-sensing practices and technologies could in one way seem to activate instrumental—or, in other words, potentially reductive and functional—approaches to citizenship and political engagement.[3] Yet, in another way, these instruments in the form of low-cost environmental sensors could rework what might be seen as instrumentalist approaches to politics to develop new vocabularies of effect and effectiveness and to challenge the apparently linear logic of these instruments through the more knotty and wayward operations of attempting to realize political change.

The seemingly straightforward practice of monitoring environments with sensors, which in turn is meant to activate political change, rarely—if ever—unfolds in such a straightforward way. Nor does citizenship magically emanate from the use of these devices. The uptake of citizen-sensing devices, then, triggers a crit-

3. For an expanded discussion and critique of these reductive and expedient sorts of instrumentalization, see Jennifer Gabrys, "Programming Environments: Environmentality and Citizen Sensing in the Smart City," *Environment and Planning D* 32, no. 1 (2014): 30–48.

ical set of questions: Which modes and practices of citizenship do these digital technologies activate, legitimate, reproduce, or transform, and who is able to operate as such a digital or sensing citizen? If the how-to aspect of these technologies is meant to guide not just the construction of sensors but also the construction of environmental citizenship, then how do sensors influence relations and responsibilities toward environments? And if the use of citizen-sensing devices does not lead to straightforward outcomes, then how do these practices instead generate open-air instrumentalisms? These questions inquire into whether citizen sensing achieves its stated outcomes and realizes its hoped-for effects, or whether different engagements unfold along the way that rework the operations and relations of these instruments. At the same time, these questions consider who gets to be recognized as a citizen and how or whether sensing technologies reproduce the inequalities that often characterize citizenship.

This investigation into *How to Do Things with Sensors* is in another way a shadowy proposition for practice-based research. Rather than undertaking an ethnographic study of citizen-sensing projects, this research works through involvement with sensor technologies, communities, environmental pollution, and political processes as a way to understand these forms of action. By working with and through practice, it is possible to query the promised effects that sensors are meant to have and to test the forms of political engagement that take hold. Yet these are also practices in the making, where technologies, citizens, and political effects are not sedimented into stable form. Practice-based research is a way to inhabit and also shape these dynamics through inquiry.[4] This research examines the practices that assemble for using citizen-

4. Jennifer Gabrys, "Citizen Sensing: Recasting Digital Ontologies through Proliferating Practices," Theorizing the Contemporary, *Cultural Anthropology* website, March 24, 2016, https://culanth.org/fieldsights /citizen-sensing-recasting-digital-ontologies-through-proliferating -practices.

sensing technologies. It considers the insights or data that these technologies generate about environmental problems. And it investigates how new forms of evidence might be mobilized to improve environmental conditions.

While these orientations might seem to point toward instrumental relations in the sense of straightforward outcomes or effects, they instead give rise to expanded understandings of instrumentalism that are more aligned with a pragmatist and practice-based approach to instruments and ideas. Practice-based engagements with citizen sensing can put these questions to the test within specific situations and in relation to distinct environmental problems. They demonstrate how technologies in practice give rise to particular ways of organizing action and generating effects that are never as straightforward as they might initially seem. These points of practice guide this investigation into citizen sensing, where tool kits have been tested and installed in numerous settings and with a wide range of participants. This how-to guide unfolds in part through recounting experiences and experiments in practice that emerged from working with environmental sensing technologies.

Getting Started: An Incomplete List of Sensor Kits

Let's look more closely at a few citizen-sensing technologies to consider how these practices are meant to present alternative strategies for documenting and acting on environmental problems. While only a few forays into the many sensors available for environmental monitoring are addressed in this how-to investigation, there is an extensive array of citizen-sensing projects that have been variously reviewed, assessed, tested, and assembled throughout the course of this research. These attempts at getting sensors up and running, developing sensor kits with communities, and working with sensor data inform this estimation on how to do things with sensors.

The focus here is on monitoring air quality. Yet there are many more sensors for monitoring water, noise, vibration, temperature, humidity, wind, heat, energy, radiation, soil, and vegetation. Water quality can be monitored through conductivity, temperature, and total dissolved solid sensors like the CaTTFish, and water levels can be monitored through ultrasonic sensors like the Flood Monitor.[5] There are Pocket Geiger sensors for measuring radiation[6] and DIY seismic sensors for measuring earthquake activity.[7] Moisture sensors monitor the health and presence of vegetation,[8] and temperature and humidity sensors monitor beehives and ensure the health of honeybees.[9] One of the earliest and most longstanding uses of DIY sensors has been in the area of monitoring air quality. Numerous sensors are now in circulation for monitoring air, from the DIY to those sold as finished products.[10] An incomplete list of operational, obsolete, and speculative citizen-sensing devices and tool kits for monitoring air quality includes

Aclima	Air Sensor Toolbox
Airbeam	AirVeda
Airbox	AirVisual
Aircasting	Alphasense Sensors
Air Quality Egg	Area Immediate Reading
AirSensa	(AIR)

5. Create Lab, "CATTfish," http://www.cmucreatelab.org/projects /Water_Quality_Monitoring/pages/CATTfish and https://www.cattfish .com/; "Flood Network," https://flood.network/.

6. Radiation Watch, "Pocket Geiger," http://www.radiation-watch.org/ and https://www.sparkfun.com/products/14209.

7. sspence, "Earthquake/Vibration Sensor," Instructables, April 28, 2016, http://www.instructables.com/id/Earthquake-Vibration-Sensor.

8. Seeed Studio, "Grove Smart Plant Care Kit for Aduino," https:// www.seeedstudio.com/Grove-Smart-Plant-Care-Kit-for-Arduino-p-2528 .html.

9. OSBeehives, "BuzzBox," https://www.osbeehives.com/; Dave Veith, "AWS IoT and Beehives," https://www.hackster.io/bees/aws-iot-and -beehives-c59fff.

10. For one extensive example of how to build an air quality sensor, see rawrdong, "How to Build a Portable, Accurate, Low Cost, Open Source Air Particle Counter," http://www.instructables.com/id/How-to-Build-a -Portable-Accurate-Low-Cost-Open-Sou.

Array of Things

Atmotube

Awair

Breathe Cam

Brizi

Cair Smart Air Quality Sensor

Citizen Sense Kit

CityAir App

Clarity

Clean Space Tag

Common Sense

DR1000 Flying Laboratory

Dustbox

DustDuino

Dylos

EarthSense Zephyr

Float_Beijin

Flow, Plume Labs

Foobot

Grove Air Quality Sensor

hackAir

IGERESS

InfluencAir

iSPEX

LaserEgg 2

LifeBasis

LondonAir App

Luftdaten

MicroPEM

Netatmo

NOKLEAD

PANDA

Plantower

Plume Air Report

PuffTrones

PurpleAir

PUWP (Portable University of Washington Particle monitor)

Safecast

SensorBox

Sensors in the Sky

Shinyei Particle Sensor

Sidepak Personal Aerosol Monitor

Smart Citizen Kit

Smoke Sense App

Soofa Benches

Speck

Tree Wi-Fi

Tzoa (Enviro-tracker)

WeatherCube

Wynd

Here are DIY sensors and off-the-shelf kits, wearables and desktop devices, as well as a few apps that double as sensors through the use of smartphones for monitoring or navigating environments. The assortment of sensor names variously suggests democratic initiatives, technical enterprises, environmental revitalization, research projects, and manga characters. The list of names further designates projects and products, communities and practices, locations and platforms. In other words, these tool kits do not de facto include certain components and exclude others. They are in varying states of composition and decomposition, salability and discontinued-ness, with different monitoring capacities.

Some of the companies or makers of the aforementioned sensors make avowedly apolitical statements, which indicates that the data the sensors collect are not intended to support political projects, nor are the data open to the collectors of air pollution data.[11] Other projects, such as the Dustbox developed through the Citizen Sense research group,[12] deliberately seek to investigate how or whether new types of political and environmental engagement can materialize with these sensors. Yet, for many of these projects, the focus is on the technical device as a way to variously organize, attract, and mobilize citizen participation and data collection. The problem of air quality is in part organized through the devices and practices that environmental sensors make possible. Monitoring air pollutants, collecting data, and communicating evidence about elevated pollutant levels could be seen to be instrumental approaches to air quality, which these instruments facilitate. The instruments and instrumentality of air quality unfold in relation to this broader proliferation of sensors and sensor citizenship. Yet these are open-air instrumentalisms that do not follow a unilateral trajectory.

In this incomplete list of citizen-sensing projects and technologies, you might notice that some of the devices are ready-made and some require assembly. Some devices are "locked down" as consumer products, while others require ongoing tinkering and maintenance. Several of the sensors included in the preceding list

11. For a commentary on how diverse forms of citizen science can align with different political (or apolitical) objectives, see Olga Kuchinskaya, "Environmental Data Collection and Citizen Science after Chernobyl and Fukushima," *Science, Technology, and Human Values* (forthcoming in the special issue "Sensors and Sensing Practices," edited by Jennifer Gabrys, Helen Pritchard, and Lara Houston).

12. For more on the Citizen Sense project, see http://www .citizensense.net/.

are beta-stage and prototype technologies that have particular idiosyncrasies and require adjusted setup, troubleshooting, and puzzling over how to work with the data that they collect and present. Over the course of this research, sensors available in makerly form have increasingly crept toward a more settled product-like state, and in the context of the Internet of Things, several plug-and-play sensors are now available. Yet, if you have worked through building and setting up air quality sensors, you are inevitably left to wonder about how or whether such off-the-shelf devices are calibrated, how to access and analyze data, and whether data can be used to support claims about environmental pollution.

This scan of air quality devices and practices is also incomplete because sensors, and especially air quality sensors, continue to multiply and expire within the usual fleeting time spans of electronic devices. As soon as sensor technologies and projects are identified, new devices emerge, and others lapse into obsolescence. Some of the air quality sensors in this list are new, some are prototypes no longer in use, and some are dead devices that would require considerable effort to reboot and plug back in to operable systems. As with many tech projects, once technologies sediment into stable forms, they seem as readily to disappear or cease functioning, with websites flickering into oblivion, firmware updates colliding with hardware configurations, and peripherals no longer communicating across ports.[13] But this observation also jumps ahead, because it points to a few things to keep in mind when starting to use citizen-sensing tool kits. First, it is worthwhile to discuss briefly how sensor kits are configured as cosmologies of sorts.

Flat-Pack Cosmology

The modular instructions and diagrams for assembling tool kits demonstrate a distinct approach to problems, where relevant

13. For a more extensive discussion of electronics and obsolescence, see Jennifer Gabrys, *Digital Rubbish: A Natural History of Electronics* (Ann Arbor: University of Michigan Press, 2011).

components are gathered together, documented, and assembled into an entity that will address the problem at hand. Think of the flat pack that consists of an itemized inventory of parts, including atomized images of assembly, with connecting actions signaled through arrows segueing across framed sequences toward a clear outcome. Similar to many modular products that can now be purchased and assembled with apparent ease, a certain flat-pack relationality is operationalized through sensor tool kits, where all the items needed for completion of the project need only be joined together by following straightforward instructions. What begins to unfold in this approach to tool kits and instructions, actions and outcomes, is a flat-pack cosmology, where the speculative universe of an environmental problem, for instance, assembles into a neat diagram of constructable relations held together through air quality sensors.

Cosmology is a term used by Whitehead to describe a metaphysical system that captures a universe of relations that is at the same time undergoing processes of transformation.[14] Cosmology also has a longer history of use within indigenous theory and practice as a term and concept that refers to sharing or holding experiences, connections, and activity in common.[15] While *flat-pack cosmology* is inevitably a heretical use of the term and concept, it indicates how the tool kit as a distributed and connected system forms and works, including the ways in which entities develop, how relations join up, how societies materialize, and how these varying components unfold and are sustained because of the values attributed to technology, especially in modular form. The difference, how-

14. On cosmology, see Alfred North Whitehead, *Process and Reality* (1929; repr., New York: Free Press, 1985).

15. See Kyle Powys Whyte, "Indigenous Women, Climate Change Impacts, and Collective Action," *Hypatia* 29, no. 3 (2014): 599–616. Within indigenous cosmology, distributions of spirituality can inform not just what Whyte calls the "instrumental value" of entities like water but also the "intrinsic value" of these entities because of their connection and agency within cosmologies.

ever, is that for Whitehead, cosmologies endure in the realm of abstraction and are drawn into the experiences of actual entities. Nevertheless, the characteristics of experiencing subjects, whether sensors or humans, express a cosmological system of relations under way. Stengers makes such a move in her multivolume text *Cosmopolitics,* where, through investigating the history and philosophy of science, she demonstrates the ways in which scientific and technical practices make particular worlds hold together, and to what effect.[16] Cosmopolitics, then, describes how these systems of technoscientific relations have political effects, and how they come down to earth.

From cosmology to cosmopolitics, Whitehead and Stengers provide reference points for engaging with the diagrams of relations that tool kits call into being. Tool kits offer up a particular way of concretizing maker-subjects, items for construction, modes of assembly, models of action, modes of becoming, desired outcomes, and ways of holding things together, or binning it all away. Tool kits are ways of organizing problems and relations for action, and even when the point is to crack open the black box of a sensor technology, the component parts that result from such crackery are configured in particular ways to allow for new modes of assembly. This taking apart and putting together, constituting and reconstituting of entities is indeed a salient characteristic of tool kits, which, on one hand, are notable for their modularity—their flat-pack-ness—and, on the other hand, need to be sufficiently open to be adaptable to new circumstances and uses.[17] Tool kits,

16. Isabelle Stengers, *Cosmopolitics I* (1997; repr., Minneapolis: University of Minnesota Press, 2010); Stengers, *Cosmopolitics II* (1997; repr., Minneapolis: University of Minnesota Press, 2011).

17. In part, I draw here on an argument made by Paul Dourish and W. Keith Edwards, who discuss how "prepackaged expectations of usage patterns" might characterize software components in tool kits, yet tool kits also need to be "designed to accommodate the wide range of potential applications and situations in use." See Dourish and Edwards, "A Tale of Two Toolkits: Relating Infrastructure and Use in Flexible CSCW Toolkits,"

especially air quality monitoring tool kits, are at once procedural and contingent arrangements. Indeed, the procedural method becomes quickly troubled in the flurry of calculating how to work with kit and in the open air of environmental monitoring. The next section details these points of procedure as well as how they go awry.

Ten Points for How to Construct Tool Kits

When starting off with a citizen-sensing project, one of the first things that many people will wonder about is how to construct a monitoring tool kit, keeping in mind that a tool kit is always more than just a collection of sensors. Some citizen-science tool kits more extensively outline techniques for learning protocols, organizing collection efforts, analyzing data, and influencing policy than the finer details of technical kit configurations.[18] Here are a few notes that outline some of the key considerations when beginning your project. After you have read these notes, we will look at a few sample projects that will give you a more detailed sense of how these points could be implemented.

1. **What is a tool kit?**
   The first thing to keep in mind when doing things with sensors is that these typically makeshift instruments will give rise to questions about the purpose, the composition, and the coherency of the technology under investigation—but never in such a philosophical way. Instead, the question will arise in the middle of attempting to get sensors to work. The refusal of electronics to function as a key part of a sensor tool kit will reveal the limited effect and scope of these devices. The eventual functioning yet occasionally inexplicable output of sensors will make one

Computer Supported Cooperative Work 9 (2000): 33–51.
18. For example, see Southwest Pennsylvania Environmental Health Project, "Citizen Science Toolkit," 2017, http://www.environmentalhealthproject.org/citizen-science-toolkit.

wonder at the apparent achievement of obtaining a connection. Sensors have not necessarily, as Simondon would suggest, become sufficiently integrated so as to seem "natural" within their own self-generated milieu.[19] Instead, they are often troublesome contraptions that would consume your time and energy as you attempt to find an operative pathway to citizen engagement. Sensors, you might discover, are just one particular entry point for engaging with air pollution, which is also interconnected with environmental public health, development, community organizing, and environmental justice.

2. **Which sensor should you use to monitor the air?**
   This is a question I am often asked by people interested in beginning to monitor with sensors. The answer is, it depends. This is the second note on how to do things with sensors. The sensor you choose to use depends on whether your interest is to tinker with electronics, to plug in a device without having to modify it, to focus on collecting "accurate" environmental data, to map and share data with a wider monitoring community, or to focus on a particular air pollutant of concern. These are not always mutually exclusive objectives, but often the focus will be placed on one priority area more than another.

3. **What are the parts you will need?**
   The third note about how to do things with sensors is that most guides will begin with a seemingly comprehensive list of parts. Photographs of parts show neatly arranged and brightly colored LEGO-like electronics that beckon for a makerly connector to join them up. The parts will include jumper cables and wire ties, Velcro and tape, breadboards and LEDs, microcontrollers and gas sensors, potentiometers and buzzers, 9-volt and lithium batteries, and resistors of various sorts. This list of parts is assembled in different ways, but the basic components include electricity and computation, held together through digital infrastructures. This is less a sensual universe of the four elements and more a functional universe of the many efficiencies. As noted, I call this how-to universe the *flat-pack cosmology*. The flat pack seemingly includes all the parts and instructions you will

---

19. Simondon, *On the Mode of Existence of Technical Objects*.

ever need to realize your objectives. However, just as your cosmology of electronics begins to assemble, you will discover that a part is missing or that the comprehensive list of parts defines sensing in one way, such as how to pass voltage along a wire, and not another, such as how to convert voltage into a semiaccurate measurement of pollution levels. The cosmology of flat pack is then always in process, splintering into multiple cosmologies of what the how-to kit might enable or open up.

4. Where should you begin?

The fourth note when working with sensors is that you are most likely better off diving in and tinkering with a bit of kit before you assiduously read the instructions or absorb extensive advice. The intricacies of pins and holes, cables and ties, are best encountered through physical proximity rather than secondhand reports. Turn to forums and videos once your brain is on the bake, the electronics refuse to talk, and you are sufficiently prepared for the curious if unique hybrid of geekery and spleen that often pours forth from makerly FAQs. Your virtual interlocutors will frequently declare, "No, I will save the planet first, and it will only be through my bespoke circuit diagram!" The how-to should, for this reason, be approached with caution in the face of such zealotry.

5. How do you make a working sensor?

The fifth note on how to do things with sensors is that the device you are working with is likely to be in need of upgrades and updates before you have even begun. The instructional guide that you are following will recommend software or hardware that will no longer be available or will be out of date. The microcontroller hardware will have been updated to a newer version. You can start with an Arduino microcontroller, to discover that the software libraries to be loaded on your Arduino no longer function with your microcontroller version. The entire configuration of the sensing kit will have shifted so that a new iteration needs to be developed through the very making and following of instructions. To do things with sensors, then, you need to trudge through states of nonconnection and electrical blank spots. Lights will refuse to flash, data will decline to post, and URLs will flash 404 where platforms should appear. The online forum and the FAQ section will become your most helpful resources in these early states.

There you will find the near-time updates and fixes that will come to your aid as you bodge your way toward a working sensor.

6. Are we there yet?
The sixth note is that a sensor tool kit will never be complete. It is a roving arrangement of stuff that will need to be topped up, updated, supplemented, extended, and hacked together. A tidy toolbox will soon become the site of a mass spill event. A clear desk will conceal an essential cable. Online warehouses will become ever-expanding otherworldly depots, where making and remaking require just one more trip to the webby aisles of Cool Components. Where does the necessary kit for undertaking a citizen-sensing project begin and end? This question could forever remain unanswered.

7. What should you do with the data?
The seventh thing to consider is that once you have built your sensor, you will need to post your data and map your monitoring locations. This process can take place on a platform that you develop along with a sensor device or on an externally developed platform that may or may not last the year. Once platforms expire, your device will likely require an entirely new configuration to pipe data to another platform. Some platforms require inputting latitude and longitude to mark fixed locations. Other platforms track the routes and itineraries of sensors used in more mobile ways. Platforms can also include the outputs of sensor data, which are presented in a wide variety of formats, from raw voltage counts to units converted to regulatory standards of parts per million (PPM) or parts per billion (PPB) or micrograms per cubic meter ($\mu g/m^3$). Pollutants monitored could include gases and particles, from nitrogen dioxide to particulate matter. These different ways of presenting and engaging with data raise multiple other how-to questions about how to analyze data sets, how to generate evidence, how to communicate findings, and how to influence policy. A successfully connected sensor, as it turns out, will be the least of your problems once you move to the domains of platforms and data.

8. How should sensors be used in the field?
Now that we're well on our way to making your first citizen-sensing tool kit, here's the eighth thing to remember: if your

plan is to monitor environments, you will eventually have to move from the desktop and workshop to the open air, where sensors will be used in situ, over time, and in a range of conditions. You might have bashed together a passable device consisting of a metal oxide sensor, breadboard, jumper cables, and resistors, all activated by a lithium battery. You might even have mapped or located your device on an online platform, thereby giving your project an apparently global reach. However, there are still yet more instructions that could be written for how to undertake pollution sensing, which necessarily extend once sensors are taken out into the open air. This guide, then, attempts to account for the contingencies, experiments, and openings that occur through such open-air instrumentalisms.

9. How do you register pollution?

Assuming that the purpose of putting together sensors is to monitor environments, then a tool kit assembled for addressing air pollution will also ideally need to address practices of environmental observation and engagement. Note 9, which would even precede note 1, is that although it can be quite easy to get bogged down in making sensor devices work, the milieus of these technologies necessarily comprise events such as how pollution registers or might register—including in existing monitoring networks, asthmatic bodies, or regulatory violations—and how communities become involved in attempting to address the problem of air quality. The scope of your tool kit might need to be redrawn so that the environmental, social, and political aspects of monitoring are as much an area of study as attempting to create a blinking LED.

10+. How do you create a community monitoring project?

Note 10 sprawling to an indefinite number of notes is that once in the open air, you will encounter endless considerations for how to make sense of your attempts to monitor air pollution. This list of further things to consider includes the following: Which air pollutants are you monitoring? Where are the likely sources of emissions? Which monitoring protocols and methods will you follow? Have you calibrated your sensor? How often and for how long will you collect data? Will your monitor be stationary or mobile? How many locations are in your monitoring network? How will you compare data across monitoring locations? How will you compare your data to other regulatory or reference monitors? In which measurement units will you present your data? How will you analyze your data? With whom will you share your data? What do you hope to change, improve, or challenge about monitored environments? With which organizations or regulatory bodies might you collaborate to act on findings from community monitoring? Think of these points as a guide that can be read alongside examples of citizen-sensing installations and in relation to your own monitoring project.

Although this section is first arranged as a ten-point plan in keeping with the exigencies of the how-go genre, it then quickly unfolds into an open-ended set of considerations when attempting to monitor pollution. As this preliminary list of overarching points related to assembling sensors for citizen monitoring demonstrates, any actual monitoring project that goes beyond the initial assembly of sensors will encounter situations that can be specific to the environments monitored. Although these tips could be presented as a set of instructions or as a checklist for how to go about monitoring, they are also far from definitive in terms of addressing the particular conditions that could arise when undertaking environmental monitoring. In the process of following these instructions, you might have learned that the assembly of sensors is far from straightforward, the composition of a tool kit is neither fixed nor complete, the posting of sensor data can come in many forms, the analysis of data is an area of ongoing development, the protocols and methods for monitoring are often still in process, the "citizens" who would monitor are often differentially able to make their voices heard, and the environments to be monitored will make specific demands upon how data are collected, presented, and turned into evidence. When following any instructional guide for sensors, you will find that open-air instrumentalisms abound.

This is not to say that it is not possible to operationalize environmental sensors for detecting pollution and gathering data. Multiple low-cost and DIY sensor projects are now in place that continually collect data and generate particular accounts of environmental processes. However, these practices are provisional and full of necessary work-arounds.[20] Inevitably, more than a few how-to guides for using sensors will present this process as a rather simple and matter-of-fact situation.[21] This how-to guide suggests, conversely, that by attending to the deviations from the straightforward approach, you could find that many more engagements with sensors, environments, and politics emerge that remake the operations of instruments and instrumentality. Open-air instrumentalisms in this way are to be valued, because they are the process through which technopolitical experiments and more just environmental collectives could coalesce. More will be said about this point in the sections that follow. Now, let's turn to look at a few detailed examples and experiences of assembling sensors for testing and eventual use in the field. A few things to ask along the way are, How does a tool kit expand or shapeshift along with differing uses? What other considerations come to the fore when attempting to monitor actual air pollution near an industry site or in a congested city? And in what ways does the how-to expand from technical delineations to indicate that these practices have been political all along?

20. For an extended discussion on the topic of work-arounds, see Lara Houston, Jennifer Gabrys, and Helen Pritchard, "Breakdown in the Smart City: Exploring Workarounds with Urban Sensing Practices and Technology," *Science, Technology, and Human Values,* advance online publication, May 26, 2019, https://doi.org/10.1177/0162243919852677.

21. One very thorough and informative guidebook that does not gloss over the many points of consideration of air quality monitoring is a citizen-sensing guidebook published by the U.S. Environmental Protection Agency. See R. Williams, Vasu Kilaru, E. Snyder, A. Kaufman, T. Dye, A. Rutter, A. Russell, and H. Hafner, *Air Sensor Guidebook,* EPA/600/R-14/159, NTIS PB2015-100610 (Washington, D.C.: U.S. Environmental Protection Agency, 2014). The guidebook is also available online at https://www.epa.gov/air-sensor-toolbox/how-use-air-sensors-air-sensor-guidebook.

How to Connect Sensors

IT COULD SEEM TO BE straightforward enough to buy a sensor, plug it in, and begin monitoring. But the situation is rarely as simple as that. The process of working with sensors includes struggles with upgrades, wonderings at data outputs on platforms, trials to test sensors in situ, and uneven comparisons with regulatory monitors. Through many practical tests of off-the-shelf citizen-sensing technologies undertaken as part of the Citizen Sense research project, it became apparent that as a user of these devices, one also became enrolled in contributing to their ongoing development by getting them to function and by contributing to (online) communities providing mutual instructions and tips for troubleshooting.

I should preface this more detailed account of how to make things with sensors by confessing that I am not, of all things, an inveterate maker. I have attempted to cobble together windmills and model cities, jelly rolls and button-down shirts, with each object bearing the sad signs of an absent hand–eye coordination. While others might have tinkered together amateur radio sets or carved out three-legged stools, for me the world of "making" has been—and remains—an ongoing but workable challenge to align bodies, forms, and functions. This how-to guide does not unfold as the advice of a seasoned expert to an audience of eager trainees (which itself is a highly gendered way in which *the digital world churns* as an ongoing performance of master-y). Rather, it is an account of dogged persistence and muddling along, of the just good enough and the bang-it-together, of the electrical tape and the makeshift arrangement. Luckily, what the hand does not or cannot make in all its supposed authenticity the computer can readily press out through a bit of code, CAD, and 3-D printing, along with numerous advice boards, collaborations, and conversations. In this sense, I engage with the world of DIY sensing as one is meant to: as an amateur connected to extended communities of practice. Indeed, it is exactly the DIY aspects of craft, making, and tinkering that can generate different experiences of embodiment, the everyday, and collective politics.[1]

In this way, I work through these faltering processes of getting sensors up and running in a citizen-sensing research group. By "up and running," I am referring not just to making photoresistors blink and microcontrollers talk but also to the extended sociopolitical and environmental relations, from communities organizing to address pollution to participatory research practices that recast the usual contours of inquiry. But this account is still not a tale of salvation through making—digital or otherwise. Instead, it is a faltering if candid encounter with the promises of DIY. It questions the processes of making kits that are meant to empower while toppling prevailing power structures. By taking up tools— specifically, DIY sensors for monitoring environments—it is possible to describe more closely how kits are made. These practices are situated ways in which to understand better the call to "hands-on" action that is meant to be a remedy for contemporary malaise. This approach involves looking at the work-arounds used to make kit operational, whether for hobbyism or environmental activism. It also describes how these technologies enable particular political engagements and ways of being and becoming citizens. Along the way, this work bypasses Heideggerian hammers to rethink and rework what counts as making along the lines of what Elizabeth Povinelli has suggested can involve a probing of differential ways of being in the world.[2]

1. Ann Cvetkovich, *Depression: A Public Feeling* (Durham, N.C.: Duke University Press, 2012). See also Helen Pritchard, Jennifer Gabrys, and Lara Houston, "Re-calibrating DIY: Testing Digital Participation across Dust Sensors, Fry Pans, and Environmental Pollution," *New Media and Society* 20, no. 12 (2018), https://doi.org/10.1177/1461444818777473.

2. Elizabeth Povinelli, "The Toxic Earth and the Collapse of Political

These accounts are a small selection of the many sensors reviewed, tested, and built. They work through the how-to, document experiences with following instructions, and elaborate on the potentials and pitfalls of working with citizen-sensing technologies. While these sensors were collaboratively assembled as part of a practice-based research process, it should also be said that some of these kits were tested and assembled in relation to pressing public events and participatory workshops in planning. In some cases, this testing and assembly process involved flying somewhat blindly into the world of sensors. These accounts re-create the steps of testing these devices after having fumbled through making and setup. In the process, I also argue that there is much to be said for using tool kits inappropriately and incorrectly. This is less a condition of embracing "error" as such and more a way of seeking out the practices that proliferate on the edges of straightforward instructions. As many feminist writers have noted in relation to their tool kits and survival guides, it is through these processes that other purportedly illegitimate or queer ways of being in the world are also forged or claimed.[3]

Making in the Imperative Mood

There are numerous guides for working with sensors, including such texts as *Getting Started with Sensors* and *Environmental Monitoring with Arduino,* among many other online manuals and tutorials. These texts could seem to be a good place to start, as they set out instructions to follow, sensors to make, and ways of toggling across making and essential concepts. The process of following one of these texts, however, yields unexpected processes of

Concepts," keynote lecture at Critical Ecologies, Goldsmiths, University of London, March 17, 2018.

  3. See Ahmed, *Living a Feminist Life*; Lauren Berlant, *Cruel Optimism* (Durham, N.C.: Duke University Press, 2011); Helen Pritchard, "The Animal Hacker" (doctoral thesis, Queen Mary University of London, 2018).

making, inquiry, and engagement. If you begin your voyage with sensors in this way, you will notice certain abiding themes for assembling technologies. Also of note with the how-to guide is that the you/your of the instructional refers to a maker who is brought into a technical relation. As I follow this instructional, I similarly traverse from first to second person to inhabit the you/your of the instructional and the imperative mood, and to work through the practice-based and participatory aspects of setting up DIY sensor technologies. Let's begin.

See, connect, attach. Insert, double-click. Orient, insert, connect, push, fold, twist, insert. Grammarians would parse this as the imperative mood. Action verbs and commands distinctly characterize the how-to of DIY electronics. Words order action. Language exhorts. Do this and complete that. Command equals outcome. Make yourself into a model citizen, and a citizen able to make models. Along the way, there will be helpful tips and conversational sidebars to review key milestones and to assure you that the makerly relationship is more friendly than authoritarian as you progress toward your goal.

The how-to guide, then, is a genre of sorts, not only in the stylistic sense but also as a way of organizing anticipation. As Lauren Berlant has noted, "genres provide an *affective expectation* of watching something unfold, whether that thing is in life or in art."[4] What is expected in the how-to guide, and how is it meant to unfold? One could say it is a genre of problem solving. Problems are identified that can be addressed through learning and sharing skills and procedures. Yet the very act of constituting problems is also a way of constituting worlds.[5] To identify the making of sensors, the writing of code, and the collection of data as key ways to skill up and undertake environmental monitoring is to commit to

4. Berlant, *Cruel Optimism,* 6.
5. See Isabelle Stengers, *Thinking with Whitehead: A Free and Wild Creation of Concepts,* trans. Michael Chase (2002; repr., Cambridge, Mass.: Harvard University Press, 2011).

and become informed by a particular way of acting on the problem of pollution. The how-to guide is meant to make acquiring and executing these practices more do-able. It also creates practices that attach citizen makers and citizen sensors to distinct ways of engaging with worlds.

In the worlds of citizen sensing, instructions are found in the form of how-to guides for making kits from assorted electronic peripherals. They also assemble as step-by-step directions for setting up and plugging in an off-the-shelf sensor to an online platform in order to view data. There are instructions for following monitoring protocols, instructions for calibration, and instructions for installing sensors in polluted locations. In the O'Reilly published text *Getting Started with Sensors,* which I detail here as the first example of attempting to work with sensors, hypothetical maker-readers begin the project of assembling sensors, while also "bending technology to [their] will" to control computation and environments.[6] Here the instructional promises a curious mastery of technology, which, as one quickly discovers, can be a bit misleading.

According to this basic primer, the process of getting started with sensors first involves asking "what are sensors?" The text readily provides the answer that a sensor is an electrical input device that "evaluate[s] a particular stimulus within the environment."[7] It would seem that any change or disturbance in an environment could be detected and transmitted into digital form. Sounding in a Simondonian register, the authors describe this as a process of "Transduction!" to explain how sensors control circuits and, by extension, environments.[8] Here environmental phenomena are undergoing conversions into electrical and digital outputs. The how prevails over the why—technology makes its own logic of execution.

Yet, when following instructions in a guide and tool kit such as

6. Kimmo Karvinen and Tero Karvinen, *Getting Started with Sensors* (Sebastopol, Calif.: Maker Media, 2014), xi.

7. Karvinen and Karvinen, 2.

8. Karvinen and Karvinen, 9.

this one, you will find that time for reflection is often cut short, since the point is to get on and make things. After asking the maker-to-be to reflect on what sensors are or might be, the authors of this instructional abruptly rejoin, "Enough discussion—it's time to build!"[9] And so you will be off, testing batteries and breadboards and LEDs, switches and alarms. There are many sensors to build here, including infrared proximity sensors, rotation sensors, photoresistors, pressure sensors, temperature sensors, and ultrasonic sensors. It will take some time to get to the finer points of connecting up air quality sensors, which are not even covered in this basic text. However, as you work through the configurations of these many sensors, you will find that it is one thing to generate a reading from a temperature sensor and quite another to know whether the sensor is providing a verifiable measurement. In the course of setting up temperature sensors in our work space on the tenth floor of an aging office building without air conditioning, for instance, we found that according to our temperature sensors, indoor summer temperatures instantly leaped to 40 degrees Celsius. Was this due to the electrical wiring of our sensor circuit, was it due to the placement of our sensor near a window on the sunny side of the building, or were we really just about to perish from heat exhaustion? Sensors at this stage of assembly can give rise to extensive questions about the state of the surrounding environment.

The temperature sensor test that this guidebook provides is to move the sensor in and out of the refrigerator, placing it in room temperature and then cooling it down in an appliance. But if you also want to speed up the process, you can expose the sensor to ice cubes, the text suggests, to obtain a more instant result. The introduction of a stimulus is often used to see whether a sensor is reading. An air pollution sensor can be exposed to a lit match, cigarette, or vacuum cleaner to see a spike or dip in the data. The first stage of connecting a sensor, then, often involves working through these

9. Karvinen and Karvinen, 4.

processes of setting up the sensor circuit configuration, loading a bit of copy-and-paste code to a microcontroller, and then testing whether the sensor detects the introduction of basic stimuli by generating detectable changes in the data.

Many of the kits that allow you to get started in testing sensors in this way are now available as defined maker kits with all the necessary parts to develop a basic plant watering system, gas sensor, or temperature sensor. Companies such as Seeed Studio sell an array of such kits that fit within the language of other assembly-based hobbies, such as model ships and airplanes. Parts to be assembled are included along with instructions, and the process of making is meant to generate new understandings of technology through doing. DIY practices on one level are meant to "challenge traditional hierarchies of authority and the existing status quo," as Matthew Ratto and Megan Boler suggest, by decentering the usual sites and practices of making.[10] Yet DIY can also reinforce particular ways of engaging with technology, for instance, as a project of following instructions to bend technology to one's will, or in other words, to gain technical mastery and to work on a universal if abstract problem. Mastery as a project has come under fire not just by philosophers of technology, such as Simondon, but also by postcolonial and decolonial thinkers, such as Julietta Singh, who suggests that a project of "unthinking mastery" can be a way to undo the estrangement that comes with these forms of relation.[11]

DIY practices can tend to be at once open-ended and instructional. What I am calling the imperative mood in this context is inevitably related to that better known theory developed by J. L. Austin of the "performative mood" in his study *How to Do Things*

10. Matthew Ratto and Megan Boler, eds., *DIY Citizenship: Critical Making and Social Media* (Cambridge, Mass.: MIT Press, 2014), 5.

11. For a critique of the master–servant relationship in technology, see Simondon, *On the Mode of Existence of Technical Objects*. For more on a discussion of decolonizing mastery, see Julietta Singh, *Unthinking Mastery: Dehumanism and Decolonial Entanglements* (Durham, N.C.: Duke University Press, 2018).

with Words.[12] The performative mood is a way of mobilizing actions, relations, and worlds through speech acts. Austin was interested in studying ways of doing things with words by considering the infelicities, misfires, and miscalculations that occur within the performative mood. His theory has in turn influenced thinkers like Judith Butler, who has further investigated how social constructs like gender are performed and materialized.[13] Karen Barad draws on and reworks Butler's discussion of performativity by adding a material and posthuman dimension that shifts discursive statements to a field of multiagential possibilities (rather than an exclusively human utterance or action). In this way, performativity is less about language in abstraction and more about what Barad calls the "conditions of mattering."[14]

Theorists of digital technologies and digital citizenship have also built on these theories of performativity to analyze how digital practices can constitute ways of performing digital citizenship.[15] While there is much more to say about performativity than space here allows, this proposal for an *imperative* mood attempts to work with and alongside this constructivist approach to words, actions, and materialities to investigate how such a mood might generate its own distinct configuration of instructions, relations, practices, technics, and milieus. Even more than drawing out the generative aspects that characterize theories of performativity, I would suggest that the misfires and miscalculations just as readily (if differently) come to the fore when engaging in the imperative mood. These

12. J. L. Austin, *How to Do Things with Words,* 2nd ed. (Cambridge, Mass.: Harvard University Press, 1975).

13. Judith Butler, *Bodies That Matter: On the Discursive Limits of "Sex"* (New York: Routledge, 1993), and Butler, *Excitable Speech: A Politics of the Performative* (New York: Routledge, 1997).

14. Karen Barad, *Meeting the Universe Halfway: Quantum Physics and the Entanglement of Matter and Meaning* (Durham, N.C.: Duke University Press, 2007).

15. Engin Isin and Evelyn Ruppert, *Being Digital Citizens* (London: Rowman and Littlefield, 2015).

misfires could also be a particular entry point into understanding how open-air instrumentalisms take hold, as swerving experiments with instruments. An instruction to insert a lead of a resistor into the same pin of a breadboard as an LED or sensor can easily end up back to front. A sensor-wire-battery configuration might connect up or become disconnected. A sensor output might be all but inexplicable and difficult to verify. Hence every list of imperative commands comes with its inevitable section on "Troubleshooting" to help makers figure out what has gone wrong along the way. But this process plays out in much more elaborate ways than simply faulty wiring or misaligned sensors. It can also extend to dodgy code and incorrect conversions as well as data platform errors and bungled sensor housing. Misfires and miscalculations can and often do extend to botched sensor installations, inscrutable data output, and indifferent responses to data gathered.

Many theorists have discussed the ways in which instruments generate more-than-descriptive engagements that enact worlds. In other words, instruments are world-making. They are constructive and performative of the worlds that they would detect, measure, and act upon.[16] But the imperative mood designates explicit actions along with observations that might be achieved. It constitutes the methods by which such constructions and performativity take place, or falter. The imperative mood constitutes the conditions, subjects, and environments in and through which a particular project is meant to occur and a particular outcome to be achieved. It tends to be normative in its register of address. When you load code onto a microcontroller, there is little sense that there are multiple ways of completing this task. Instead, you pursue the project in the seemingly correct way, which you are meant to master to move on to the next step. This configuration of technology, maker-citizen-subject, technical relations, and practical action is what the imperative mood designates.

16. For example, see Donna J. Haraway, *Modest_Witness@Second_ Millenium.FemaleMan©_Meets_OncoMouse™* (New York: Routledge, 1997).

This initial example of working with sensors vis-à-vis a standard maker text such as an O'Reilly guidebook is meant to demonstrate the particular entry point and process whereby sensor instruction occurs. Sensor ontologies quickly give way to flat-pack cosmologies, where component parts join up to create electrical arrangements of action and reaction. The progression through a guidebook, and formation of your own tool kit, can be delineated as the working through of devices: from LEDs on to temperature and pressure sensors. In other words, guidebooks can tend to direct makers to a particular notion of technical mastery without considering the open air where devices would not only be installed but also shapeshift in their broader arrangements. When making citizen-sensing tool kits and following instructions for assembling sensors, it is important to consider how these instructions are organizing a particular way of encountering the problem of monitoring environments as well as establishing technopolitical relations.

Simply Connect

Once you've attempted to assemble sensors from their basic component parts, you might next decide to test a sensor that is off-the-shelf and does not require an intricate process of assembly with breadboards and jumper cables. As noted, an increasing array of sensor objects and off-the-shelf products can be procured through Kickstarter pledges and online shopping. The Air Quality Egg, which I detail here as the second example of attempting to work with sensors, is perhaps one of the most iconic of these citizen-sensing devices. Having been prototyped through a series of hacker events from 2010 to 2012 (which I describe in *Program Earth*[17]), the Egg moved from prototype to saleable product by 2013, when the Citizen Sense research project was under way. Because the Egg was an air pollution sensor that could be purchased as a complete product, it did

17. See Jennifer Gabrys, "Sensing Air and Creaturing Data," in *Program Earth*, 157–81.

not require soldering or microcontroller setup. Such an off-the-shelf sensor would seem to allow for more attention to be given to environmental monitoring, data collection, and public engagement, and so our research group began to investigate the capacities of this device.

We placed an order for our own Egg in the middle of July 2013. The sensor arrived at our offices in London in late July, sent from Wicked Device, based in Ithaca, New York. The neatly packaged device promised on its outer label to make more engaged citizens of us all. It declared,

> Problem Solved. Do you ever think about the air you breathe? It affects us in ways we can see and also in ways we can't. The Air Quality Egg is a project working to make the air we breathe more "visible." Simply hang it in your home, office or outside your window to start collecting your personal air quality data. The Air Quality Egg connects you with a global community of concerned citizens participating in the ongoing air quality conversation.

The strangely daunting prospect of simply plugging in and connecting to a global community of concerned citizens able to solve an environmental problem as intractable as air pollution meant the Egg actually sat on our shelf for another week. Opening the kit seemed to be a ceremonial event for which we best waited and prepared.

So when, in early August, we set aside time to unbox and install the Egg, we made a wager as to how much time would pass before we were successfully gathering air quality sensor data. Estimates spanned from having the device up and running within the afternoon to a few days. A more skeptical researcher estimated that it could take months, if ever, until sensor data were coursing through this plastic Egg and transforming into environmental solutions. You might find yourself engaged in such speculation as to what the final setup and output from your off-the-shelf sensors could be. This process is central to the ways in which citizen-sensing practices take form as ongoing contingencies of devices, environments, and engagement.

Once we had unpackaged the Air Quality Egg and scanned the different components of the kit, we next read through the seemingly straightforward how-to instructions. We began the setup by entering the device's serial number (or MAC address) into the Air Quality Egg Google map platform, where we also located and named our Egg.[18] With this, we were able to see the Citizen Sense Egg in Southeast London, situated within a wider and global community of sensing citizens, albeit one that numbered around 250 in population worldwide. Here was an apparently eager if niche community of Egg owners, ready to solve the problem of global air pollution. Once our Egg was on the map, we turned to setting up the device and posting data. Yet you might find, as did we, that putting your sensor on a map is often the most basic and straightforward of the technical challenges that you will encounter with the Egg.

The Air Quality Egg is formed of a pair of translucent white plastic Eggs: a "base" station Egg that at the time of testing posted data to the Xively platform and a "remote" Nanode Egg that does the job of sensing air pollutants and gathering data via a shield outfitted with metal oxide nitrogen dioxide and carbon monoxide sensors and with temperature and humidity sensors attached to an Arduino microcontroller. Although there have been updates and a new version of the Egg since the time of this testing, with data now posted to a different platform and updates made in sensor setup, this was the device arrangement with which we worked at the time. If you refer to points 5 and 7 above, you'll find that the Air Quality Egg presents numerous dilemmas in the form of upgrades and fixes. These events are not exceptional to the Egg, because not only are these relatively new and unstable devices but also they inevitably succumb to the rapid rates of obsolescence that are characteristic of electronics.

During Egg setup, we next discovered that the power plug was configured for U.S. electrics. As we did not have a power converter, we had to set the project aside until we sourced an adapter for the U.K. context. This was a simple enough prob-

18. "Air Quality Eggs," https://airqualityegg.wickeddevice.com/portal.

lem, but we found that this was just the beginning of several stages at which we realized additional kit would be needed to make the Egg function. Once we eventually powered the device, we found that it did not perform the correct color sequence to indicate that it was collecting and posting data. Flashing color sequences were the means by which we were to be made aware of air pollution through the successful posting of environmental data to the platform. But the sequence of our flashing colors did not follow the same sequence outlined in the setup instructions. We spent some time trying to determine what the color conversions were indicating, exactly, when we discovered that the color issues were due to a bug, discovered several months earlier in January 2013 but still impacting devices shipped as late as ours, where the Eggs were no longer talking to the Xively platform for displaying data. An elaborate process then ensued of attempting to reprogram the Egg base and remote Nanode, following the official Air Quality Egg Google forum and FAQ instructions, which linked us to a seemingly straightforward video indicating how to fork a repository of code from GitHub to reprogram the Arduino microcontroller in the Air Quality Egg.

And yet, after completing the process of reprogramming, the Eggs were still not producing "data," neither in the form of flashing color leading to awareness nor in the form of line graphs on Xively or cryptic bar charts on the Google Maps Air Quality Egg page. Eventually, through multiple waves of turning the device on and off and reloading and verifying code, we managed to obtain blips of data, strange right angles in line graphs, and numbers apparently in parts per billion of nitrogen dioxide or carbon monoxide but generally remaining inexplicable in what measurement they presented, exactly. We had a modest assurance that the temperature and humidity readings might be somewhat correct, because these agreed with other sensors we had in operation in the same space, but converting the nitrogen dioxide and carbon monoxide measurements to a legible figure was a rather more difficult matter, as we did not have other verifiable sensors in operation for comparison.[19] Our closest point of reference was with the official London Air Quality Network station, and here the measurements were not in units that could be easily cross-referenced.

By getting the Egg up and running, we had experienced what one of our developer-collaborators called the "'Hello World' of IoT." Getting an air pollution sensor setup to post data was a basic achievement along the pathway of the Internet of Things. Yet, in the process of connecting the Egg, we were more intent on asking this IoT "world" a whole host of other questions about how these devices were sensing air pollutants. What were these sensors sensing, exactly? How could we find out more about the hardware and software setup and the extent to which this could influence the data outputs? How did the "color-equals-awareness" engagement with environmental data work, exactly, especially when color did not signal pollution levels?[20] At the same time, what sorts of data were these, in terms of their accuracy, legibility, and legitimacy, when even getting a device to work, whatever the readings, seemed to be an achievement? And now that our device was operational, what were the capacities of the network of concerned citizens to which we were connected? From our use of the Air Quality Egg forums, it seemed that the communities we connected to had more interest in hobby electronics and computer tinkering than mobilizing their data to influence air quality policy or enforcement.[21] Indeed, the Egg was now circulating in the world in ways

19. For a related discussion of atmospheric science practices that attempt to generate legible data, see Emma Garnett, "Developing a Feeling for Error: Practices of Monitoring and Modeling Air Pollution Data," *Big Data and Society* 3, no. 2 (2016): 1–12.

20. For a discussion of these inscrutable aspects of the Air Quality Egg, see Christian Nold, "Device Studies of Participatory Sensing: Ontological Politics and Design Interventions" (doctoral thesis, University College London, 2017).

21. For an in-depth ethnography of the communities involved with making and testing the Air Quality Egg, see Dorien Zandbergen, "'We Are Sensemakers': The (Anti-)Politics of Smart City Co-creation," *Public Culture* 29, no. 3 (2017): 539–62.

where maker communities effectively contributed free R&D by testing the device, while finding fixes and improvements through necessary troubleshooting.[22]

The universal citizen sensor, embedded in a community of global citizens, then comes down to earth by having to work with the specificities of particular infrastructural configurations. The seamless plug-and-play logic and practices that such devices would promise continually meet with simple obstacles and more complex malfunctions. The imperative mood here might instruct you to plug in your Air Quality Egg and connect to other global and instrumental citizens. But your inability to complete the command or follow the instructions can multiply into a whole set of other practices, infrastructures, and relations that you will likely need to call on to fulfill seemingly straightforward instructions. In a similar way, Lucy Suchman has pointed out that even a task as obvious as pushing a green button on a photocopier machine can give rise to confusion, adjustment practices, technical communities, and modified technological artifacts that materialize when simple button-pushing does not yield the expected results.[23] Instructions seem to guide a technical encounter, but they do not determine it. Instead, the "sense" made of and through instructions materializes through the actual undertaking of a technical practice.[24] Practice-based research in this way is an approach that surfaces the many adjustments and deviations that can arise when working

22. This process might align with what Suchman and others, drawing on Garfinkel, have discussed as co-constituted courses of "instructed action" in relation to the prototype. With the prototype, the "configuring" of devices and actions, working practices and sociomaterial relations, is one that relays "across sites of technology development and use." See Lucy Suchman, Randall Trigg, and Jeanette Blomberg, "Working Artefacts: Ethnomethods of the Prototype," *British Journal of Sociology* 53, no. 2 (2002): 168.

23. Lucy Suchman, *Human–Machine Reconfigurations: Plans and Situated Actions,* 2nd ed. (Cambridge: Cambridge University Press, 2007), 8–9.

24. Suchman, 22.

with technologies in lived situations. Instruments and instrumentalities frequently deviate from simple action and outcome. These are the misfires of the imperative mood, which generate the open-air instrumentalisms of citizen-sensing technologies.

You might then find that off-the-shelf sensors promise that you can "simply connect"—and by extension also connect to greater air pollution awareness and a community of global and instrumental citizens—and yet they do not unfold in such a straightforward or liberatory fashion. Instead, they generate open-air instrumentalisms that deviate from the process of following instructions and setting up tool kits. Air quality sensors do not always immediately function, either technically or in terms of their broader sociopolitical and environmental effects, but in the malfunction of devices and the reconstitution of instructions, other worlds-in-the-making are generated along with technological subjects. The misfires that percolate through the imperative mood occur in part through anomalous technical arrangements that come into being, as detailed here, and in part through the ways in which these devices circulate and are taken up to address environmental problems. While these sensors proved to be anything but off-the-shelf, there is the possibility (if not the danger) that the promise of such modular and ready-to-use devices also could begin to inhabit the space of politics, encounters, and relations, for instance, in the form of off-the-shelf politics, off-the-shelf citizenship, and off-the-shelf public engagement. This is why it is important to ask what sorts of instrumentalisms are mobilized with apparently ready-made sensing technologies and sensing practices.

At the time of this writing in 2019, the Air Quality Egg has undergone many updates and is now available for sale in newer versions.[25] The website notes that "big improvements" have been made over

<hr/>

25. Wicked Device, "Air Quality Egg," https://shop.wickeddevice.com/product-category/air-quality-egg.

the version 1 Egg that we had tested. Indeed, its supplier, Wicked Device, no longer supports version 1, and Xively, the platform host, has terminated the data service that the version 1 Egg used. To post data, the Egg would need to be reconfigured to post to a different data platform. As the Wicked Device announcement states about this option, "That's a fair amount of work, and will require that you recompile and re-load your Egg with the new service destination."[26] The easier route is to purchase the version 2 Air Quality Egg, which guarantees an even more seamless plug-and-play experience. A new device replaces the defunct one, and the promise of technical action becoming democratic action is refreshed.

Whereas the Air Quality Egg on one level seems to promise that air quality monitoring can become a relatively effortless affair, many plug-and-play sensors require considerable effort to become operational. Updated and upgraded versions will still require ongoing maintenance and fixes as well as skilling up to learn about technical configurations. Even with all of this effort, there are still many more questions about the verifiability of the data that plug-and-play devices generate as well as the protocols and practices that are used when monitoring environments. While this is in no way meant to deter you from testing out citizen-sensing technologies, this how-to setup that involves working across standard instructions as well as actual practices undertaken is meant to demonstrate the misfires and miscalculations that proliferate when inhabiting the imperative mood and when working with the genre of the how-to and the tool kit.

Around the time we had undertaken our own provisional setup with the Egg, I began to be asked by representatives from local governments, environmental NGOs, and even air quality officers in small nations whether they could replace their expensive air quality monitoring instruments and networks with Eggs. My cautious reply

26. Dirk Swart, "Egg Version One End of Life," May 19, 2017, https://shop.wickeddevice.com/2017/05/19/egg-version-one-end-of-life.

was, not unless you'd like to spend considerable time dealing with misfires and miscalculations. While many of these off-the-shelf devices can be and are used in interesting ways that add to the scope of DIY practices, they are tetchy gadgets that produce a variable range of data that currently do not transfer well to the spaces of air quality regulation. While a modest achievement can be made in getting a sensor device such as an Egg up and running, its flickering displays and data outputs do not necessarily sync well with the expanded technical, social, political, and environmental requirements of air quality governance in its usual sense. For this, you might need to engage with even more versions of the how-to, including points 7 to 10 given earlier, which indicate how the technoscientific configuration of an air quality sensor and the data it generates depend upon extended infrastructures to make sense.[27] These are infrastructures not just of technical capability but also of stabilizing data-as-evidence to address the experience and event of air pollution.

The Reluctant Prototype

Parallel to, and perhaps even in advance of, working with the Air Quality Egg, the Citizen Sense research group was in the process

27. Latour demonstrates the ways in which the how-to, as a form of instruction, can proliferate through scientific infrastructures that travel along with artifacts to ensure that they are suitably encountered. For instance, in relation to a natural history museum collection, he writes, "Even those elements which can withstand the trip, like fossils, rocks or skeletons, may become meaningless once in the basement of the few museums that are being built in the centres, because not enough context is attached to them. Thus, many inventions have to be made to enhance the mobility, stability and combinability of collected items. Many instructions are to be given to those sent around the world on how to stuff animals, how to dry up plants, how to label all specimens, how to name them, how to pin down butterflies, how to paint drawings of the animals and trees no one can yet bring back or domesticate." See Bruno Latour, *Science in Action: How to Follow Scientists and Engineers through Society* (Cambridge, Mass.: Harvard University Press, 1987), 225.

of developing prototype air quality sensing kits of our own making, which I detail here as the third example of attempting to work with sensors. These were provisional groupings of nitrogen dioxide and particulate matter sensors, which we had used in a pilot walk in the New Cross area in London in early July of that same year.[28] You might find that as you progress from following O'Reilly and Instructable tutorials, your own devices assemble neither as makerly stuff nor as off-the-shelf kit but as particular prototypes that are cobbled together in a cut-and-paste and makeshift way. A sensor configuration that works in one setup can be morphed over to another expanded kit, and code passes along on these various iterations or is drawn from libraries to create a new workable concoction.

In just this way, we were attempting to put together a possible prototype citizen-sensing kit that might be used as a tool of engagement while undertaking fieldwork in the United States, as we were researching fracking in Pennsylvania.[29] Multiple citizen-monitoring activities had already emerged to address pollution and public health concerns taking place in Pennsylvania. In this context, we wondered what role a prototype kit could play in engaging people to ask questions about environmental data, how the data are generated, and their effects in addressing the problem of air pollution. How might data be shared and collectivized? How might they travel differently than the current if complex modes of

28. Jennifer Gabrys, "Air Walk: Monitoring Pollution and Experimenting with Forms of Participation," in *Walking through Social Research,* ed. Charlotte Bates and Alex Rhys-Taylor, 144–61 (London: Routledge, 2017).

29. For a more extended discussion of the "Pollution Sensing" project area of Citizen Sense, see Helen Pritchard and Jennifer Gabrys, "From Citizen Sensing to Collective Monitoring: Working through the Perceptive and Affective Problematics of Environmental Pollution," *Geohumanities* 2, no. 2 (2016): 354–71, and Jennifer Gabrys, Helen Pritchard, Nerea Calvillo, Tom Keene, and Nick Shapiro, "Becoming Civic: Fracking, Air Pollution and Environmental Sensing Technologies," in *Civic Media: Technology, Design, Practice,* edited by Eric Gordon and Paul Mihailidis (Cambridge, Mass.: MIT Press, 2016), 435–40.

reporting on well locations and pollution levels? In the process of building a kit comprising multiple air quality sensors, including nitrogen dioxide, carbon monoxide, temperature, humidity, and particulate matter 2.5 sensors, we found that as many questions were raised about the validity of data that might be generated from such a kit and the possible promises and expectations that could be raised by circulating environmental technologies to communities affected by pollution from fracking. We were beginning to engage with sensors in the open air, not just by moving them to actual sites of pollution detection, but also by collaboratively testing them with communities knowledgeable about documenting pollution through environmental data collection and analysis. In this context, open-air instrumentalisms multiplied and abounded even further.

This tool kit in the making then demanded that we think through the instructions and how-to pointers that we might provide so as to make the kit legible and usable. Questions that came up when thinking about how these kits might be used in the field included, Would there be a manual with instructions for use? Would each sensor be explained in relation to what it senses and how chemical detection optimally works? How long would the sensors need to operate to collect usable data? How long should the sample rates and duration of monitoring be? Would the sensors work only if stationary, and should instructions indicate to keep the kit stable during use? If we are recording data, would these be made available to individual participants, or would they be shared collectively on a web platform? Should these data be given locational information or be made anonymous? What instructions might participants need to analyze the data in order for them to be meaningful? Would the sensor housing skew the readings in any way? What are the base readings of the sensors, and are we sure they are properly calibrated? Would the sensors or pollutants interfere with each other? Could we be sure they are sensing exactly what they are meant to sense? Would the kit be damaged in shipping from London to Pennsylvania, and what adjustments might need to be made in the field?

These questions were also informed by an attempt to ensure that multiple participants' engagement with the kit might be collaborative and experimental from the beginning and not only a functional end application. At this point in the development of sensors, we queried the notion that by collecting data—accurate, skewed, or otherwise—environmental politics would be more readily democratized or facilitated. This was an attempt to critically engage through practice with a seemingly more instrumental–functionalist agenda. Yet while we sought to critically examine the role of environmental-monitoring technologies in forming practices and politics, rather than simply becoming advocates of this approach, we found that we also had to take seriously the instrumental logics of these devices and the citizen-sensing practices they activated and organized. These reworkings of instrumentality became part of the way we experimented with making alternative citizen-sensing tool kits that could engage with the practices and concerns of participants (who were engaged in preexisting monitoring projects), and generate usable environmental data, but that would also open into other engagements with environmental problems.

So, with all of this in mind, in autumn 2013, we began the process of making prototypes to test the ways in which sensors generate, influence, and operationalize environmental data. Our version 1 Citizen Sense Kit initially consisted of two primary devices, one a sensor shield pulled from the Air Quality Egg, which included nitrogen dioxide, carbon monoxide, temperature, and humidity sensors that we attached to a combination of a Grove Board to add a real-time clock and a Raspberry Pi microcontroller. The version 2 Citizen Sense Kit comprised stand-alone sensors (rather than a pluggable sensor shield), including carbon monoxide, nitrogen dioxide, particulate matter 2.5, temperature, and humidity sensors, where we added our own resistance configurations that we found considerably affected the readings in comparison to the Egg shield.

With both of the preliminary Citizen Sense Kits, the first intention was first to get the devices up and running, because in the process of making the kit, even more questions emerged about the how of the how-to. The second intention was to disassemble and reverse-engineer more black-boxed technologies, such as the Egg, which on one level required all sorts of capacities and resources to make function and on another level had rather unclear information about how the hardware and software were put together, how the sensors were configured, what resistance was used and how this affected data outputs, and how continual changes of the data platform "back end" (from Pachube to Cosmo to Xively) could affect the data's form and analysis. You might find yourself asking similar questions if you attempt (or have attempted) to make sensor tool kits using devices sourced and assembled through multiple configurations. Repurposing and questioning become key techniques in assembling these kits.

In this way, the Citizen Sense Kits were built through information from multiple forums, as there was no official single forum from which we might obtain guidance on how to make the monitoring technology "work." We developed even more iterations of the Citizen Sense Kit, including a version 3 that included a Speck particulate matter sensor, an analog BTEX (or benzene, toluene, ethylbenzene, and xylene badge), a Frackbox (for monitoring volatile organic compounds and nitrogen oxides), a monitoring logbook, and a data platform. This version of the kit was later used to monitor pollutants from fracking at thirty community locations from autumn 2014 to early summer 2015.[30] Along the way, we foraged for diagrams and work-arounds; forked code from GitHub; and shoveled piles of breadboards, resistors, and cables across desktops. By rebuilding kits, and gakining another perspective of the hardware and sensor configurations, we were also able to observe along the way what technical resources, capacities, and infrastructures these technologies require as well as the decisions

30. Citizen Sense, "Citizen Sense Kit," https://citizensense.net/kits /citizensense-kit.

that were made or elided to make the monitoring kits in these particular ways, and the domains inhabited to generate and circulate environmental data through these contraptions.

By working with air pollution–sensing tool kits, we tested how environmental sensing technologies enable certain types of monitoring and generate questions about the limits and possibilities of each of these monitoring practices for addressing environmental and political problems. The practice of making prototypes and setting up off-the-shelf sensors becomes a way to work through the instructions, promises, functions, and malfunctions of these devices. It also generates open-air instrumentalisms. In the process of procuring air quality sensor guides, making kits, following instructions, and installing devices in the open air, a number of splintering pathways came into view by deviating from a straightforward approach to these devices. Online forums read as tales of ongoing struggles to set up sensors, to maintain their operations, and to update and adjust when upgrades are available. FAQ sections are brimming with queries about connections, data, and modifications. Platforms bear the traces of half-finished efforts in running sensors, where maps of monitoring locations click out to nonexistent line graphs or inexplicable charts. These open-air instrumentalisms began to take on a more-than-technical quality as sensors were readied for installation and use, where an initial success at connection splintered into multiple concerns about the use or relevance of these devices.

But this is not to say that devices never arrive at a condition of organized use or implementation. Instead, it is to signal the ways in which setting up citizen-sensing technologies is an ongoing trial, a back-and-forth effort of testing and tweaking. At the same time, despite the democratic selling points, many devices remain tied to practices focused primarily on technology and "making" and so can become somewhat self-referential in their pursuits, thereby missing the promise to address—and even improve—environmental problems. Yet if, as Dewey has suggested, the "invention of new agencies and instruments create[s] new ends," then how do these new instruments "create new consequences" and "stir" us to "form new purposes?"[31] This is a question about instruments and instrumentality, which the next section considers in a more reflective key.

31. John Dewey, *Logic: The Theory of Inquiry* (New York: Henry Holt), 78.

How to Devise Instruments

Every science must devise its own instruments.

—WHITEHEAD, *Process and Reality*

WHEN WHITEHEAD ASSERTS THAT "every science must devise its own instruments,"[1] he is referring in part to the need for distinct tools to be formed in relation to modes of inquiry. A study of ecology materializes as a much different inquiry than a study of philosophy. Here Whitehead notes, "The tool required for philosophy is language."[2] His statement has a multidirectional character and suggests not only that tools are required for distinct scientific practices but also that scientific practices are formed through devising and using distinct instruments. While in addition to philosophy, this assertion could point to physics and mathematics, biology and atmospheric chemistry, it might also indicate how the practices of citizen science and citizen sensing form with and through distinct instruments and instrumental processes. Yet are instruments also defining entities for these practices, which might variously be characterized as what Ruha Benjamin has called a "people's science"?[3] And if so, how are these instruments further characterized by their modes of practice and not just by their distinct form as tools? In other words, citizen science and citizen sensing cross the spectrum of possible tools and subjects of inquiry, yet it

1. Whitehead, *Process and Reality*, 11
2. Whitehead.
3. Ruha Benjamin, *People's Science: Bodies and Rights on the Stem Cell Frontier* (Stanford, Calif.: Stanford University Press, 2013).

is not just the actual instrument used that is the defining characteristic but also the mode of engagement and relationality set in
motion that remakes instruments, scientific practice, and inquiry.
This section considers, then, how instruments materialize along
with practices of inquiry and inquiring subjects—the instrumental
citizens that would undertake sensing projects.

Instruments do work in the world. They can make undetectable
phenomena evident. They tune us into other registers of experience, and they attach us to perceptive practices that remake our
sensory worlds. A list of instruments devised along with scientific
practices could extend to epic proportions, spanning the fantastic
and the precise. If the air pump has featured to demonstrate the
emergence of a certain mode of objective science,[4] it has also been
the source of much attention in producing universalized subjects
who are seemingly detached from making their objects of inquiry and knowledge.[5] In this way, instruments and machines have
served as devices for differentiating the contours of a rational
human subject from an automaton or a duck, for instance.[6] In a
contrary way, technical devices, such as engines, are seen not to
be mere instruments and instead have been discussed as generative of new subjects, milieus, and relations.[7] Instruments might
also seem to be something distinct from the contours of the human
body, but as writings on the cyborg have demonstrated, instruments can remake technologies, subjects, relations, environments,
and politics as well as what counts as scientific inquiry.[8]

---

4. Steven Shapin and Simon Schaffer, *Leviathan and the Air Pump:
Hobbes, Boyle, and the Experimental Life* (Princeton, N.J.: Princeton
University Press, 1985).

5. Haraway, *Modest_Witness,* 23–45.

6. René Descartes, *Discourse on Method and Related Writings,* trans.
Desmond M. Clarke (1637; London: Penguin, 2003).

7. Simondon, *On the Mode of Existence of Technical Objects.*

8. Donna Haraway, *Simians, Cyborgs, and Women: The Reinvention of
Nature* (New York: Routledge, 1991); Ian Hacking, "Canguilhem amid the
Cyborgs," *Economy and Society* 27, no. 2–3 (1998): 202–16.

Environmental sensors as they are used within citizen-sensing practices are similarly wide ranging. Hygrometers and anemometers, barometers and thermometers, as well as metal oxide and electrochemical sensors for detecting air pollutants: there is a roving tool kit of borrowed, appropriated, hacked, and repurposed parts that citizens work with to attempt to document environmental disturbance. Here are multiple instruments, with different tunings, standards of measurement, modes of observation, political effects, and world-making capacities. If instruments are integral to the practice and definition of what counts as science, then you might wonder how citizen science and citizen sensing, with their DIY and makeshift instruments, begin to challenge and rework not just instruments but also what counts as science. What are the instruments of citizen sensing? How would citizen sensing be variously defined in relation to those sensors listed earlier, from PuffTrones to the Air Quality Egg? How do these devices contribute to the formation of diverse practices of inquiry? What are their capacities for transmogrifying the evident to make new forms of evidence? These are questions to ask along the way, while wondering about these instruments-in-the-making.

Instruments are a long-standing topic of investigation in science and technology studies, and there is no shortage of investigation of scientific instruments.[9] Rather than trace the historical lineages and social–epistemic formations of instruments, however, I am interested in investigating the uneven and sprawling ways in which contemporary citizen-sensing instruments are taken up to pursue environmental and political agendas, but also how they at times fail to realize these outcomes or arrive at different alignments than

9. For example, see Lorraine Daston and Peter Galison, *Objectivity* (Brooklyn, N.Y.: Zone Books, 2007); Andrew Pickering, *The Mangle of Practice: Time, Agency, and Science* (Chicago: University of Chicago Press, 1995); Cyrus Mody, *Instrumental Community: Probe Microscopy and the Path to Nanotechnology* (Cambridge, Mass.: MIT Press, 2011); Liba Taub, "Introduction: Reengaging with Instruments," *Isis* 102, no. 4 (2011): 689–96.

initially anticipated. This is a way of working within, while also reworking, instruments and instrumentality toward something more like open-air instrumentalisms. The instruments of citizen sensing demonstrate how apparently instrumentalist versions of evidence-based politics can give rise to diverse and inventive citizen-based and collective practices through the very attempt to gain influence through the collection of data. These practices not only complicate an easy critique or adoption of instrumentalism; they also reinvent relations with instruments and instrumentality. At the same time, instruments or tools are already mutually constituted with practices, so that new worlds concretize through engagement with instruments, but not as a linear process.

Instruments are invariably involved with social relations. Any change to them, as Latour has suggested, will also shift social conditions. As he writes, "change the instruments, and you will change the entire social theory that goes with them." Here Latour is engaging with the work of Gabriel Tarde to note the way in which "science is *in* and *of* the world it studies," where instruments become crucial to social relations as they are performed, lived, and understood.[10] A change of instruments, along with the standardization of instrumental processes, also in-forms the worlds that are set in motion and sustained. There is an instrumental force that instruments contribute to operationalizing. Yet these ends, consequences, and purposes are less likely to materialize through the mere enactment of a script or program built into an instrument and more likely to concretize through the social worlds and political subjects that assemble along with and through instrumental processes. You might wonder if there is also a how-to aspect to Latour's assessment of Tarde. In other words, how do you change the instruments so that you can also change the entire social theory that goes with them? In its search to devise instruments, the

10. Bruno Latour, "Tarde's Idea of Quantification," in *The Social after Gabriel Tarde: Debates and Assessments,* ed. Matei Candea, 145–62 (London: Routledge, 2009).

how-to guide could be a call to undertake experimental engagements that generate ways of working with and through new technical arrangements, infrastructures, and modes of governance. The how-to guide is not simply the study of a technical problem; it is also an encounter with the potential of other social worlds. When thinking about how to devise instruments, you might then consider how changing instruments also changes the possibilities of encounter, engagement, and relation.

Instruments and Instrumentality

Instruments are the tools, devices, and contraptions that are constituted as they do work in the world. An instrument can be a sensor, a data logger, a tool kit. There are also conceptual instruments, discursive instruments, and policy instruments. An instrument could be something that standardizes and measures but also that constructs and generates. Instruments and instrumentality, then, are processes that put modes of inquiry and experimentation into motion, while raising questions about the types of observation and action that are undertaken. Are instruments generative of an expanded instrumentality, or are they prescriptive in their engagements and outcomes?

Instruments are often described as "mere" or "passive" or "functional" devices. Simondon suggests that an instrument-based view of technology tends to be reductive. He writes that the technological object has been "treated as an instrument" considered in relation to economics, work, or consumption but not engaged with through philosophical or cultural deliberations.[11] Unlike cultural objects, he suggests that technical objects are relegated to "a utility function" and do not enjoy "citizenship in a world of significations."[12] For Simondon, the designation of a technical object

---

11. Simondon, *On the Mode of Existence of Technical Objects*, xii–xiii.
12. Simondon, 16.

as an instrument is a way of focusing on its functions only, where instrumentality seemingly has a predetermined outcome: to complete the task at hand. By suggesting that this is a way of denying technical objects a sort of citizenship, he seeks to diversify the entities from and through which meaning and sense—meaning as sense—materialize. Tools and technics, in other words, are cultural relations and expressions.

However, while in Simondon's analysis, an instrument might be seemingly fixed in its capacities and modes of observation or operation as well as outcomes, it is also subject to retooling. As Whitehead notes, language is not simply a tool used by philosophy; instead, "philosophy redesigns language in the same way that, in a physical science, pre-existing appliances are redesigned."[13] This redesign occurs in part because of the breakdown of that instrument, which occurs at the edges "of expressing in explicit form the larger generalities—the very generalities which metaphysics seeks to express."[14] As instruments are engaged in processes of inquiry, they are then worked and reworked as they approach the edges of inquiry. But breakdown and retooling are not the only conditions of this instrumental engagement. These conditions occur because instruments—in this case, language—are searching toward propositions of fact that are also referring to the universe needed to sustain those facts.[15]

Here we are again in the thick of a flat-pack cosmology, but from another perspective. The universe is not ready-made from a tool kit, nor is a tool kit as ready-made as it might have seemed to be. Instead, a universe is required for instruments to be put to work, making both the tools and the universe somewhat indeterminate in the inquiry to be undertaken, because they are both in process. An instrument might reach toward something more fixed and ab-

13. Whitehead, *Process and Reality*, 11.
14. Whitehead.
15. I discuss the environments needed to sustain facts in a related register in Gabrys, "Sensing Air and Creaturing Data," 157–81.

solute, as it will require its world to make sense, yet these are both in the making through the practice of inquiry.[16] The instruments and instrumentality that might have seemed to project toward a certain outcome become open-air instrumentalisms. They form through relations and deviate from a fixed purpose. They take shape through distinct modes of inquiry.

The many components of scientific practice involve what Jenny Reardon and collaborators refer to as the "material relationships that are part of knowledge-making practices, including political, social and cultural ones."[17] The instrument that might be accounted for in a scientific investigation is then always connected to "the multiplicity of entangled apparatuses" that includes ethics and justice.[18] While the focus could easily lead to human-makers taking up instrument-tool kits to address environmental problems, such a perspective would further demonstrate the ways in which instruments, observation, observer, and phenomena are entangled such that world-making is a distributed and multiagential affair. In this sense, apparatuses for Barad "are constituted through particular practices that are perpetually open to rearrangements, rearticulations, and other reworkings."[19] Resurfacing here is a certain breakdown and redesign—or retooling—that occurs not as the work of a willful human subject but rather as part of the shifting conditions in which instruments unfold through instrumental operations and relations. Open-air instrumentalisms are multiagential and not only are the work of makers or tools but also erupt through situations, practices, and relations.

In the process of making instruments, you might wonder wheth-

16. This is what Whitehead refers to as the "impossibility of tearing a proposition from its systematic context in the actual world." See *Process and Reality*, 11.

17. Jenny Reardon, Jacob Metcalf, Martha Kenney, and Karen Barad, "Science and Justice: The Trouble and the Promise," *Catalyst: Feminism, Theory, Technoscience* 1, no. 1 (2015): 13.

18. Reardon et al.

19. Barad, *Meeting the Universe Halfway*, 170.

er your approach to sensing environments has become somewhat "instrumental" or, as usually designated, overly functional. But as this discussion begins to suggest, even that which seems to be defined as an instrument and its instrumental outcomes begins to break down and be retooled through practices of inquiry. Although instrumentality has acquired a negative connotation—to say that something is instrumental is to suggest that it is a grossly efficient means to an end—these critiques of a certain mode of causality deserve another look in the context of working with citizen-sensing instruments.[20] Although citizen-sensing technologies are often wrapped in the promise of a simple means–end practice, where sensing the environment will generate political change, the instrumental operations of these instruments are never as simple as this. Instrumentality can demonstrate other modes of effect and effectiveness, not that of a reductive cause and effect but rather a multiagential making of worlds.[21] Although instrumentality might seem to generate a limited set of engagements with problems, this revisiting of instrumentality from within the milieus of instruments-in-practice shows how other modalities of action and practice can materialize.

Instrumentality is a mode of experience that might be productive of particular observations, expressions of citizenship, and relations with other collective entities for acting on problems of environmental pollution and environmental harm. Instrumentality, in this sense, necessarily becomes experimental in the process of undertaking concrete action. *Instrumental experimentalism* was a term and concept that Dewey used in a somewhat interchangeable

20. For a discussion of critiques of instrumental reason and instrumental control through the works of Heidegger and Habermas, see Andrew Feenberg, *Questioning Technology* (Abingdon, U.K.: Routledge, 1999). Also extending the Heideggerian consideration of technology and instrumentality, Hannah Arendt takes up this topic in relation to her conception of "Homo Faber" in *The Human Condition* (1958; repr., Chicago: University of Chicago Press, 1998).

21. Barad, *Meeting the Universe Halfway*.

way with *pragmatism* to refer to the contingency of "ends" within a philosophical—or democratic—project.[22] On one level, Dewey was accounting for the rational unfolding of concepts that is central to the pragmatists' approach to the instrumental. He drew on Charles Sanders Peirce and James to elaborate on a general progression of logical concepts as they are taken up in concrete situations. On another level, Dewey indicated the ways in which instrumentalism had implications for democracy—as a conceptual project always likely to generate struggle, contested relations, and modes of governance that are not direct or effortless instantiations of democratic principles.[23] Or as Cornel West has suggested, such a "future-oriented instrumentalism," while on one hand ran the risk of heroic or individual approaches to creative democracy, was on the other hand a search for strategies of "more effective action."[24]

A propositional end might serve as a guide for concrete action, but it is always provisional and inevitably reworked through concrete experience and practice. Because an end is not merely arrived at, moreover, it is in many ways radical in relation to the instrumental experimentalisms it operationalizes but from which it also deviates. Instrumentalism for Dewey, then, is about a process of experimentation, inquiry, and discovery. In this sense, it would be

22. As Dewey further writes, "instrumentalism is an attempt to establish a precise logical theory of concepts, of judgments and inferences in their various forms, by considering primarily how thought functions in the experimental determinations of future consequences." Writing also about the work of James, Dewey suggests that the "reconstructive or mediative function ascribed to reason" becomes a way to develop "a theory of the general forms of conception and reasoning." This suggests that the experimental processes of instrumental concepts are then the means by which theories cohere into general forms rather than instrumental approaches proving a priori truths. Instrumentalism in this rendering is necessarily experimental and contingent. See Dewey, "Development of American Pragmatism," 14.

23. See John Dewey, *The Public and Its Problems: An Essay in Political Inquiry* (University Park: Pennsylvania State University Press, 2012).

24. West, *American Evasion of Philosophy*, 5.

possible to say that instrumentalism has always been experimental. Moving from logical concepts to those other instruments, the material technologies that would also guarantee a certain rational unfolding of machine logic in the world, we find even more prospective instrumentalities under way.

In this respect (and in contrast to Simondon), the instruments of citizen sensing are not instrumental enough, since they seem to guarantee an outcome that would foreclose the very undertaking of citizen-sensing practices as concrete experiences. This a priori designation of an outcome reduces not just the instruments and instrumentality but also the instrumental citizens that would materialize through these practices. Although the terminology is different, with this Simondon might agree: the conventional promises of citizen sensing constrict instruments into functional outcomes, a process that forecloses inquiry or experimentation. Here Simondon might be inclined to admit instrumentalism to his analytical tool kit, because this does the work of reclaiming the processes of inquiry and open-endedness that he suggested were more appropriate to understanding and transforming human relations with technology. Shutting down and narrowing inquiry, as Dewey suggests, limits the modes of experience and political engagement that could be possible.

The instrumentalism developed here takes a cue from these pragmatist approaches to experimentation and inquiry and is informed by the open air that James found was necessary to practices of inquiry. "A pragmatist," as James writes, "turns away from abstraction and insufficiency, from verbal solutions, from bad *a priori* reasons, from fixed principles, closed systems, and pretended absolutes and origins." Instead, pragmatism involves a turn "toward concreteness and adequacy, towards facts, towards action and towards power."[25] While this is a steer toward a certain kind of (radical) empiricism

---

25. William James, *Pragmatism and Other Writings* (New York: Penguin Books, 2000), 27.

to which James was partial, it also directs inquiry toward "the open air and possibilities of nature, as against dogma, artificiality, and the pretence of finality in truth."[26] *Open air,* as I develop the concept with and beyond James, refers to an operationalized and prospective approach to inquiry. *Open air* pertains to lived experience, to processes of inquiry as they are unfolding, rather than to doctrines to which inquiry is made to conform. Dewey, expanding on this aspect of James's work, suggests that instruments, or ideas, become "true *instrumentally*" through the ways in which they "work." The working aspects of ideas were far more relevant than the final outcome that might seem to offer up a resolution, such as truth. At the same time, instruments are also "a program for more work, particularly as an indication of the ways in which existing realities may be *changed.*"[27] Open-air instruments and instrumentalisms, then, are tool kits for practice; they are able to generate change, above and beyond a static pronouncement of truth.

Expanding on James and Dewey, I move from the unfolding of logical instruments to practices with technical instruments to suggest that instruments such as citizen-sensing technologies are more than a means to an end. As it turns out, it is only by undertaking practices and engagements with and through instruments that contingent relations and capacities begin to materialize, demonstrating that instrumentality has never been quite so straightforward as it might have seemed.[28] This is the scope of open-air instrumental-

26. James.

27. James, 28–30.

28. Ian Hacking notes that instrumentalism came to suggest a certain "antirealism" within the philosophy of science. As West has pointed out, however, pragmatists such as a Dewey, James, and Peirce worked with realist ontologies, while evading the fundamental epistemological concerns of philosophies concerned with truth. This sidestepping of epistemologies of truth does not make pragmatism antirealist; rather, in West's estimation, it contributes to the evasion of more Cartesian concerns with knowing how the real is really real through a preconceived division of subjects and objects. See Hacking, *Representing and Intervening: Introductory Topics in the Philosophy of Natural Science* (Cambridge: Cambridge University Press, 1983), and West, *American Evasion of Philosophy.*

isms: to demonstrate how sociotechnical practices are set to work and how they make and change worlds. While Dewey sought to clear up the confusion about the terminology and meaning of instrumentalism, I work with this productive dissonance to query the trajectories and outcomes of sensing instruments. I propose the term *open-air instrumentalisms* as a way to capture this revisiting and reworking of instrumentality within the context of DIY environmental sensors. This concept and term is about more than logical propositions, as it also captures the prospective qualities of instruments and instrumentalities.

The open air, then, punctures any closed logic of instrumentalisms. Despite the imperative directions of guidebooks and tool kits that would suggest a quick passage from flat-pack cosmology to actionable gadget, you will find there is not a simple way to bend technology to your will. Instead, here are the tool kit, the instructional, the guidebook, and the instrument unfolding into the open air of instrumental experimentalism. Rather than instrumental reason giving way to a singular means–end trajectory, these are open-air instrumentalisms that, when put to work in the world as practice, concrete experience, and contingency, engage with and generate multiple inhabitations. Instruments not only contribute to organizing inquiry in particular ways; they also distribute inquiry across multiple entities and relations, creating new communities of inquiry—something I will address later. Instruments are involved in tuning and in-forming environments, worlds, and political subjects that further trouble the usual scope of instrumentality: those instrumental citizens.

Instrumental Citizens

By drawing on multiple and diverging thinkers, I expand on the notion of what instruments and instrumentality might mean or generate. There are many different uses of the term *instrument* across these thinkers, and they are by no means synonymous. The instruments of Simondon are merely functional technologies; the instruments of James and Dewey are theories and ideas put to work in the world; the instruments of Whitehead become part of practices of inquiry; and the instruments of Barad expand out to relational, material, and entangled apparatuses. If Whitehead's remark at the beginning of this section has much to say about science and instruments, it says less about the subjects caught up in these instrumental practices. Who or what are these instrumental entities? If "citizens" are monitoring environments with sensing instruments, do they become instrumental citizens? Are they instrumentalized in the conventional or in the pragmatist sense? And do they realize new political competencies through their instrumentalist practices, which theorists of feminist technoscience and indigenous and critical race theory develop as strategies of retooling technologies?

If Dewey's and James's sense of the instrumental were applied to the citizen, it might mean that the democratic commitments of political subjects would always be put to work, and that it is through this work that the very meaning of *citizen* would come to have consistency. The work of citizens, then, continues to generate the reality and community of citizens as well as the transformed instruments that would further spur this work along.[29] Instrumental citizens in this sense are not rationalized actors completing a designated task—the reductive or functional sense of *instrumental*; rather, they are contingent subjects involved in making and remaking—tooling and retooling—political life. Indeed, as ongoing work in environmental justice has demonstrated, the retooling of instruments and devices occurs along with the transformation of politics and relations in order to work toward less polluting environments.[30]

29. For a related discussion on the "work" of political, collective, and democratic life, see Helen Pritchard and Jennifer Gabrys, "From Citizen Sensing to Collective Monitoring: Working through the Perceptive and Affective Problematics of Environmental Pollution," *Geohumanities* 2, no. 2 (2016): 354–71.

30. For example, see Jason Corburn, *Street Science: Community Knowledge and Environmental Health Justice* (Cambridge, Mass.: MIT Press,

A citizen-sensing kit as much comprises citizens as sensors. Yet the "citizen" is not an entity that can be wired and coded in the same way as a microcontroller. Instead, drawing on Simondon, we could say that what the citizen is or could become is "in-formed" by sensors as well as the extended milieus in which they operate. In this sense, we have already begun to recompose the citizen-sensing tool kit even before beginning a process of assembly. The "citizen" in these citizen-sensing tool kits is meant to be an action-based entity. This is a citizen that is imagined to be an empowered and effective technophile. The instrumented citizen is an instrumental citizen, in the usual sense of realizing a stated outcome through direct and efficient action.

We might further question the seeming logic and expediency of sensing instruments and instrumental citizens to effect change, which is a narrative that depends on leaving many aspects of technical and political engagement unquestioned. In this sense, it is worth pausing for a moment to examine in more detail the usual diagram of how citizen-sensing action is meant to unfold and how the designations and expressions of citizens and citizenship are meant to be performed through sensing technologies. For instance, Plume Labs, which has developed a wearable Flow sensor as well as an AI-powered app for forecasting air pollution, focuses on the ways in which citizens as sensors might monitor their own air to protect themselves and their families from high pollution levels.[31] While they suggest that collective undertakings are possible, Plume manages and oversees the collected and collective data in such a way that they are not readily available for use and analysis by communities. In this sense, personal action and protection are emphasized and collective action is deferred into a process and space defined

2005), and Carla May Dhillon, "Using Citizen Science in Environmental Justice: Participation and Decision-Making in a Southern California Waste Facility Siting Conflict," *Local Environment* 22, no. 12 (2017): 1479–96.

31. See Plume Labs, "Clean Air, Together—Flow by Plume Labs," September 26, 2017, https://www.youtube.com/watch?v=uJtahMBOn6g.

by the technology company. This arrangement might assure user-consumers that by monitoring their air, they are not sliding into the dangerous depths of citizen activism but rather are maintaining a more neutral engagement with technology to protect themselves and their families. The site of engagement becomes a more nuclear and normative undertaking less inclined to the sprawling affiliations of democratic communities beyond the family.[32]

The configuration of subjects is one of vulnerable and responsible family members managing their personal air space in a politically neutral manner. The air here becomes more like a Sloterdijkian atmosphere of air conditioning and security—a space of instrumental control rather than open-air instrumentalisms.[33] Citizen-sensing instruments are enrolled to facilitate this management, thereby shoring up a particular citizen-as-consumer engagement with the problem of air pollution. This is far from an isolated example of how many consumer-based air pollution technologies are now being promoted, whether in the European Union or the United States, China or India, the focus is on managing and protecting oneself and one's family members in controlled personal spaces. Awareness, especially personal awareness, is the way in which instruments and instrumentality are organized. Yet this raises the question of what awareness is meant to spark

32. For analysis of these different ways of parsing the state, the community, and the citizen through or beyond the family, see Lauren Berlant, *The Queen of America Goes to Washington City: Essays on Sex and Citizenship* (Durham, N.C.: Duke University Press, 1997).

33. A popular reference for discussing air pollution and air control, Sloterdijk's work nevertheless strikes an essentialist and deterministic note in its rendering of the air as a space of terror and control. This study deliberately sidesteps this more fixed reading of air as an "element" "essential" for life, not least of which because of the rigid political imaginaries that issue forth along with these atmospheric ontologies. See Peter Sloterdijk, *Terror from the Air* (Cambridge, Mass.: MIT Press, 2009), and Jan-Werner Müller, "Behind the New German Right," *New York Review of Books,* April 14, 2016, http://www.nybooks.com/daily/2016/04/14/behind-new-german -right-afd.

into being. These "aware" subjects are not directed to intervene in current operating conditions to undertake democratic struggle toward more breathable collective atmospheres. They are instead made aware so as to better manage their own individual exposure. Of note here is that citizen-sensing technologies for monitoring air pollution are increasingly shifting away from DIY and maker-ly technologies toward finished consumer products. Instruments in this context do not as readily give rise to a Deweyan process of experimentation and inquiry. Instead, they potentially direct consumer-users to a series of corrective or adaptive actions not dissimilar to a cybernetic logic that Simondon critiqued for its functional approach to technical objects, which overlooked the ways in which technologies undergo processes of concretization.[34]

Here the citizen also becomes utility-like in the imaginings of citizen-sensing technologies—able to singularly and instrumentally effect change. But the reflections of Dewey suggest that we might consider other forms of instrumentality in relation to politics. Following the pragmatists, Antonia Majaca and Luciana Parisi also suggest that instrumentality is not instrumental, at least not in the way it is usually conceived. In their estimation, through a more thorough engagement with the logic of *technē*, it might be possible to go about "reversing the very understanding of instrumentality," which could be undertaken "by fully acknowledging instrumentality, politicizing it, and ultimately transcending it."[35] In their estimation, transcending instrumentality entails recognizing that subjects are also contingent, and this contingency is where the political materializes through concrete practices.[36] With this understanding of the instrumental formation of political subjects, instrumental citizens become entities that are in formation and are involved in world-making activities. Because subjects materialize

---

34. Simondon, *On the Mode of Existence of Technical Objects*, 51.
35. Antonia Majaca and Luciana Parisi, "The Incomputable and Instrumental Possibility," *E-Flux Journal* 77 (November 2016).
36. Majaca and Parisi, 1–3

through modes of instrumentality, they cannot be fixed into absolute categories. The specific hold of instrumentality in a given situation involves the working through of a prospective engagement, which is the formation of the political. A further elaboration upon the concept of an instrumental citizen would then involve taking up Dewey's notion of "instrumental experimentalism" as the putting to work of what a political subject is or could be. This is less a fixed mode of engagement and more an opening into how the citizen as attractor and force can stir us to new purposes, as previously discussed.

While Dewey opted to use *instrumentalism* as a term interchangeable with *pragmatism,* he also worked with *experimentalism* as another term and concept that attempted to explain and capture the ideas he was developing. In science and technology studies, experiments and experimentality are frequently discussed to describe the way in which these open-ended practices of inquiry and engagement take place.[37] These are not technical solutions but rather provisional practices that generate new approaches to technologies and new engagements with politics. Hence the relevance of this discussion for understanding what a citizen is and might become. The instrumental is at first seemingly similar to the experimental, because it is a contingent and open-ended process. However, it is different in that it is a practice guided by ideas, technologies, and tool kits that seek to do a certain amount of work, and even possibly (political) transformation, in the world.

37. Drawing on Dewey, Ana Delgado and Blanca Callén investigate DIY biology and electronic waste hacking experiments to consider how "hacks" as an "experimental mode of inquiry" open up new approaches to problems. See Delgado and Callén, "Do-It-Yourself Biology and Electronic Waste Hacking: A Politics of Demonstration in Precarious Times," *Public Understanding of Science* 26, no. 2 (2017): 179–94. See also Javier Lezaun, Noortje Marres, and Manuel Tironi, "Experiments in Participation," in *The Handbook of Science and Technology Studies,* 4th ed., ed. Ulrike Felt, Rayvon Fouché, Clark A. Miller, and Laurel Smith-Doerr, 195–222 (Cambridge, Mass.: MIT Press, 2016).

Instrumentalism is not a test for the sake of a test or an experiment for the sake of an experiment. The fact that instruments and instrumentalities are unlikely to fulfill their stated aim is not a limit but rather crucial to the process of working out instruments and ideas for further development. As James writes when discussing the work of Dewey, *"theories thus become instruments, not answers to enigmas, in which we can rest."*[38] This may explain why Dewey used the term *instrumentalism* as well as *instrumental experimentalism* to describe this putting to work of instruments.[39] *Experimental* describes the contingent and open-ended modes of action, but instruments are the things and concepts put to work and reconfigured through experimental processes.[40]

The point of revisiting and reworking instrumentalism as open-air instrumentalisms is not to recuperate a reductive notion of technical or political action but instead to consider how neither citizens nor machines have ever been instrumental in the usual sense of the word. The adoption of a citizen-sensing instrument does not make for a more direct realization of a citizen-scientific or citizen-political impact. Instead, it organizes modes of inquiry, social relations, facts, and worlds. Instrumentality, then, describes the instrumental commitments that are taken up in distinct practices as well as what these generate in the open air. The logic of sensing

38. James, *Pragmatism and Other Writings*, 28 (emphasis original).
39. Dewey, "Development of American Pragmatism," 20.
40. As West has pointed out, it is worth noting the specific ways in which experimentalism emerged in pragmatist thought, where the scientific method was seen to be a paragon of "critical intelligence" and experimentalism was very much a product of this practice—and where the "social base" for such pragmatism required a more elite professional class to engage in such practices. Nevertheless, West suggests a possibility for "creative democracy" might still persist in relation to experimentalism. See West, *American Evasion of Philosophy*, 62, 90, 97, 103. There are more and other expanded ways of engaging with experimentalism—and instrumentalism—that might also be developed beyond the registers of pragmatism and the scientific method, which are the topic of the expanded and forthcoming version of this work.

that is used to promote these technologies as a direct solution to environmental pollution could be understood as a form of instrumental reason that diminishes a more contingent and experimental understanding of instruments and instrumentalities as well as the practices of scientific and political inquiry. Instrumental reason is bound to bend, instruments will unfold through contingent operations, political projects and struggles will become activated and entangled with instrumental experimentalism.

When taken up and put to work, the instruments and instrumentalities of citizen-sensing technologies break down, open up, and are retooled through particular practices and communities. Instrumental citizens in this sense are political subjects (which are not necessarily always human subjects) that are working through the problem of sensing environmental pollution to make more livable worlds. Here it might be possible to suggest that technical objects could be granted "citizenship in a world of significations," as has previously been discussed through Simondon. However, such citizenship is never settled but is instead undertaken through the differential and multidirectional practices of human and nonhuman instrumental citizens as they sense, rework, and retool tool kits and their worlds.

How to Build Networks

THE MODES OF ACTION delineated through citizen-sensing technologies, including making and coding, monitoring and data collection, are expressed with and through instruments. The command to "get practical" as well as the exhortation "enough discussion—it's time to build!" are calls to action that tool kits and guidebooks, the how-to, and the imperative mood organize and deliver. Yet these instrumental imperatives take on a much different meaning once we have reworked and retooled instruments and instrumentalities toward open-air instrumentalisms.

By building and getting "practical," a shift in current operating conditions is meant to occur. Such instruments and instrumentalities demonstrate how technoscientific practices, instruments, subjects, and worlds are collectively generated, along with an estimation of what the consequences of these instrument-worlds might be. The "practical" is what James refers to as "the distinctively concrete, the individual, the particular and effective as opposed to the abstract, general and inert."[1] Yet for James, when expanding from pragmatism to radical empiricism, the distinctively concrete refers not merely to things but also to relations.[2] Or as Haraway has suggested in her discussion of yet another instrument, the air pump, "nothing comes without its world."[3] You might find that "getting practical," then, requires a greater engagement with the sprawling relations, networks, and worlds that materialize along with instruments.

1. James, cited in Dewey, "Development of American Pragmatism," 6.
2. William James, *Essays in Radical Empiricism* (1912; repr., Lincoln: University of Nebraska Press, 1996).
3. Haraway, *Modest_Witness*, 37.

Constructing tool kits and connecting sensors are practices that further expand into techniques for building networks. Getting practical is always an encounter with and formation of relations. The setting up of one device moves from making or plugging in a sensor and piping data to a platform to connecting with and comparing data across multiple sensor nodes. But this computational approach to networks is only one way of configuring what a network is or might be as it concretizes through citizen-sensing technologies. While it might at first have seemed the primary focus, when taken into the open air, a sensor becomes one small component within a broader project of addressing environmental pollution. Indeed, when it comes to monitoring air quality as an environmental problem, communities are often already mobilized in various ways to document and address pollution. Networks are in the making, but they do not start from zero. Sensors become part of the practice of community organizing, and technical relations transform in the process.

Because networks are already at work in the world, Citizen Sense set out to learn from and alongside existing environmental-monitoring practices. This is a method of first learning about who is monitoring, where, and why. During our research, which involved online searches, attending community meetings, arranging interviews, making site visits, distributing logbooks, hosting mapping workshops, and guiding monitoring walks, we found that communities were monitoring air, water, noise, and traffic by using analog and digital sensors, including particulate sensors, air pollution badges, decibel meters, FLIR infrared cameras, video and photography, and CCTV installations. Communities were also using professional lab testing services, gathering and consulting planning documents, keeping track of changing land surveys, and monitoring policy and regulation as well as petitioning for changes and improvements to environmental controls.

In the course of organizing to address environmental pollution, identifying pollution sources and tools to monitor emissions becomes just one aspect of different modes of inquiry and action. Communities work with existing networks for organizing environmental projects, and they find ways to contribute to and build on these in relation to concerns about environmental pollution. They also contact regulators and policy makers to register complaints about pollution, host community meetings, gather evidence about health conditions, give public testimony, share news on social media, set up teleconferences, contact experts and public figures to extend and amplify networks, and document pollution with assorted sensing technologies. Instrumentalities shift here, where to "build" something involves much more than making a digital device operational. A project to monitor and address air pollution involves building community monitoring networks as ongoing, iterative, and contingent practices that are ways of making and maintaining technical, social, political, and environmental infrastructures.

Perhaps somewhat different from citizen science, citizen sensing has a more specific focus on digital tool kits and devices, so that these organizational, collective, and environmental aspects of monitoring might at first seem to be of lesser importance. However, in this way of configuring what a community monitoring network is or might become, it is clear that sensor tool kits develop into much more than digital gadgets or makerly components. By working with situated environmental problems, citizen-sensing practices and technologies quickly become bound up with wider networks of environments, communities, institutions, and politics. The accuracy of monitoring devices, the monitoring protocols used, the legitimacy of the data, and the agendas of users all come into play as factors influencing the techniques of environmental monitoring and the data gathered. Citizen-sensing practices move from the more reductive diagram put forward by the Air Quality Egg to shift instead into distinct networks of inquiry and political contestation. In the process of making sensors, you might then find that these technologies proliferate along with different types of networks that include the communities of inquiry that make, install, query, and operationalize citizen-sensing technologies.

Communities of Inquiry

The practice of building a community monitoring network involves building and drawing on communities of inquiry. The process of taking an instrument into the open air does not merely consist of testing or setting up a device. Instead, a tool kit develops along with networks and inquiries. Community of inquiry is a concept that Peirce developed to describe scientific modes of inquiry and how reality, facts, and truth are settled on through collective processes.[4] This phrase was in turn taken up by other pragmatists, such as Dewey, to describe the way in which concrete practices of inquiry generate realities that are particular to groups undertaking such work. For Dewey, these modes of inquiry become political, informing the possibilities and struggles of democratic life.[5]

The how-to can involve multiple processes of inquiry. But this is not merely an abstract set of instructions followed by a universal subject. Instead, the how-to as inquiry is situated within communities. Along with these formations of communities, inquiry, and facts, it is clear that instruments are also put to work as they transform along with communities of inquiry.[6] This is an expansion of a

---

4. Charles Sanders Peirce is generally credited with having developed the notion of community of inquiry. Peirce's notion was developed in relation to the pursuit of logic and science, but pragmatists have adapted the concept, especially Dewey in relation to democratic modes of inquiry. As Peirce writes, "unless we make ourselves hermits, we shall necessarily influence each other's opinions; so that the problem becomes how to fix belief, not in the individual merely, but in the community." See Peirce, "The Fixation of Belief," in *Chance, Love, and Logic: Philosophical Essays*, 7–31 (1877; repr., Lincoln: University of Nebraska Press, 1998).

5. Dewey, *The Public and Its Problems*.

6. Working within a different context, Grant Wythoff describes how "communities of amateur tinkerers" experiment with technologies and gadgets to become the "engine of emerging media." Experimental inquiry in this more hands-on sense becomes part of the process whereby technologies further develop and concretize. See Grant Wythoff, ed., *The Perversity of Things: Hugo Gernsback on Media, Tinkering, and Scientifiction* (Minneapolis: University of Minnesota Press, 2016), 37.

community of inquiry that includes nonhumans in their technical and fleshy arrangements and instantiations. This approach to the how-to process becomes decidedly less about a maker tinkering with a digital object and more about the collective constitution of worlds. How-to is a way of collectively organizing and asking how to go about something, including how to make a world. How are communities of inquiry organized in relation to environmental problems? What are their practices, tactics, and strategies? But the question does not merely document the occurrence of networks; rather, it also contributes to the prospective formation of networks. This is part of what you might attend to when making a citizen-sensing tool kit.

As a prospective undertaking, inquiry, then, is a mode of transformation. Another approach to the how-to emerges here, where how indicates or asks in what manner, by which means, and how might it be possible to organize ways of life. How indicates procedure, practice, and process. The imperative mood shifts to become less commanding and more aligned with a certain obligation and necessity. In her text *Imperatives to Re-imagine the Planet*, Gayatri Chakravorty Spivak outlines a mode of the imperative that involves recasting the relations of subjects through planetary connections that exceed that which can be designated or made commensurate with subjects.[7] The imperative in this sense is as much an opening as a responsibility, a proposition as well as a commitment to justice. How attends to the mode of engagement and the imperative of attending to what is at stake in attending to and attempting to address (if not redress) planetary troubles.

The how-to, then, opens up to engage with another register of the imperative: the crucial actions undertaken that contribute to lived engagements that remake worlds. Elaborating on this aspect of the how, indigenous theorist and writer Leanne Betasamosake Simpson writes in relation to Michi Saagiig Nishnaabeg thought and practice,

7. Gayatri Chakravorty Spivak, *Imperatives to Re-imagine the Planet* (Vienna: Passagen, 1999).

"It became clear to me that *how* we live, *how* we organize, *how* we engage in the world—the process—not only frames the outcome, it is the transformation. *How* molds and then gives birth to the present. The *how* changes us. *How* is the theoretical intervention."[8] How is the world-making process that traverses ways of life, modes of politics, registers of experience, and integrities of relation. It forms subjects and environments in its indication toward engagements. It is both a theoretical orientation and an embodied collective practice.

Expanding Practices of the How-To

When using sensors in the open air, the question of how to assemble tool kits then expands into other orders of instrumentality and the how-to. Is this just a matter of distributing air pollution sensors to a community? Or is this very process a way of forming new networks and communities of inquiry? And if it is the latter, then how might it be possible to expand upon the makerly way of encountering sensors to engage with these devices as more fully social technologies that are constituted in and through diverse more-than-makerly social environments? How-to is a question that frames a problem, and that indicates how to act on that problem.[9] The how-to of citizen-sensing tool kits frames the problem of environmental pollution as one of measurable quantities that can be documented and communicated as evidence. Yet this how-to also organizes an

8. As Simpson further writes, "engagement changes us because it constructs a different world within which we live." See Leanne Betasamosake Simpson, *As We Have Always Done: Indigenous Freedom through Radical Resistance* (Minneapolis: University of Minnesota Press, 2017), 19–20.

9. These questions also introduce what science and technology studies scholars have referred to as the "politics of how." See Endre Dányi, "The Politics of 'How,'" *Does STS Have Problems?* (blog), October 4, 2016, https://stsproblems.wordpress.com/2016/10/04/the-politics-of-how, and John Law and Solveig Joks, "Indigeneity, Science, and Difference: Notes on the Politics of How," *Science, Technology, and Human Values* 44, no. 3 (2018): 424–47.

expanded set of practices, from how to build a community monitoring network that responds to the sited problem of pollution to how to draw on community expertise and connections and how to gather observations and experiences of environments over time.

As one example of an approach to exploring this expanded configuration of the how-to, Citizen Sense built upon its ongoing practices of meeting with community groups and residents concerned about air quality by developing a "Logbook of Monitoring Practices." The logbook sought to find another way of assembling the how-to, through collaborative research and development. The logbook was one part of the Citizen Sense Kit that organized techniques to constructively and collectively ask about the how: how to build a tool kit, how to use sensors, how to monitor, how to use data, and how to effect improvements to environments and environmental pollution. This was a process that did not start from a preformed assumption about what technology is or ought to be but rather asked what it could become within a community of inquiry committed to collective engagement with environmental problems. You might find that by asking questions about the how-to with a more low-tech device, such as a logbook, it can be possible to configure an expanded tool kit through a process of collective research.

In this way, our first "Logbook of Monitoring Practices" was developed as a series of questions to ask participants about how they would document the problem of air pollution from fracking. These questions were entry points into the how-to: they asked how to go about monitoring this complex and fraught environmental problem that people had struggled with for years. These questions could be worthwhile to consider when developing your own tool kit that also seeks to build community monitoring networks. The questions include the following.

1. What pollutants should be monitored?

From benzene and carbon monoxide to particulate matter, nitrogen oxides, ozone, light, and noise, a number of possible pollutants are associated with the industrial process of fracking.

This set of questions also asked what the main pollutants of concern were and what the tool kit should include to document these pollutants.

2. Where should monitoring take place?
Pollution can occur throughout the hydraulic fracturing process and across its infrastructure, including at drill sites, well pads, compressor stations, glycol dehydrators, impoundment ponds, and pipelines. Here the logbook asks where the most noticeable emission sources are. It also asks what possible overlooked pollution sources might not be monitored or regulated.

3. Who is monitoring?
Some monitoring activities might already exist and could be undertaken by government agencies or industry. This set of questions asks who might be monitoring already, who should be monitoring if this is not taking place, and in what way the data should be accessible.

4. What monitoring practices are citizens already undertaking?
When pollution is suspected to be occurring, it is common for nearby residents to begin monitoring to determine whether their air, water, soil, and surrounding environment could be causing harm. By learning more about existing practices, it is possible to incorporate these knowledges and experiences within the development of expanded tool kits.

5. What exposures have been noticed or felt?
By asking about how exposures are experienced, as well as the distance between natural gas infrastructure and homes, it can be possible to understand the health effects that could be linked to emissions.

6. What is difficult to monitor or can't be recorded?
In the fracking process, there are a number of unknown substances in drilling fluids, surfactants, slurry, lubricants, and foaming agents. This question asks about uncertainty about environmental pollution as well as the possible limits of monitoring equipment for detecting different substances.

7. How should citizen data be used?
Sensors can generate considerable amounts of new data, and when this is multiplied across a community monitoring network,

large data sets can accumulate. This set of questions asks how these data could be used, what effect they might have, and whether and how the data should be shared across the community or farther afield.

8. What does a day in the life with fracking look like?

By asking participants to document what everyday life with pollution involves, it can be possible to pick up the many activities that could be causing pollution as well as associated effects. Everyday life might also have shifted in relation to the ongoing operations of the industry, and this question captures observed changes to environments over time.

9. What monitoring scenarios should be tested?

When monitoring different infrastructure, distinct monitoring setups could be useful to investigate. For instance, it could be worthwhile to monitor the "life of a well" as it is graded, spudded, drilled, and finished as a well pad producing gas. Monitoring might also take place at set distances from the emissions source. This open-ended question asks participants to consider what a monitoring experiment could look like, to develop a research design, and to put it into action.

10. What additional observations can be added?

Because residents observe changes to environments over time, and witness the effects of pollution as they take hold, you might find it is useful to ask for photographic documentation of environmental changes as well as any additional observations or questions that can inform the how-to of the tool kit.

These logbook questions for composing a monitoring tool kit are less an absolute list to follow and replicate and more a provisional map of the ways in which different questions—that ask how to rather than instruct how to—can assemble a process of inquiry, a monitoring tool kit, a set of environmental observations, an indication of how to work with data and evidence, and an understanding of community networks and interests. Participants' contributions included lists of pollutants to monitor, including particulate matter, volatile organic compounds, methane, nitrogen oxides, and noise. These were important starting points for how we then came to build tool kits to be installed near infrastructure. In the logbooks, clear indications emerged of what parts of the fracking infrastructure were of particular concern, including compressor stations and well pads. New information surfaced that might have been overlooked about problems with traffic, including industry trucks, heavy equipment, and helicopters, all serving as the moving infrastructure for hauling fracking materials and waste to and from sites. Logbook contributions also offered detailed suggestions for who should monitor and monitoring scenarios, including to encircle industry sites with monitors and provide real-time data to the public.

The range of environmental events, changes, and pollution that participants added to the logbooks became a complex if informative record for considering how to make a relevant citizen-sensing tool kit. As one participant documenting "a day in the life with fracking" noted, their experience of fracking was characterized by

Trucks, trucks, and more trucks.

Traffic tie-ups *much* more frequent.

Dust blowing everywhere.

Hills crisscrossed with pipelines.

Slow super heavy equipment on oversized trucks.

60' wide swaths of trees coming down to make roads.

Country lanes being widened and built up completely changing the character.

Bright lights in the sky at night, near and far.

More helicopter traffic.

Noise from drilling, trucks, flaring and compressor stations at all hours.

More litter on the roadways.

Torn up roads.

Torn down barns.

Not safe to ride a bike on the backroads anymore due to trucks barreling around curves.

No more rhythm to life—no downtime. No weekends or holidays. Industrial intrusion 24-7-365.

Neighbors uncomfortable at best, fighting with each other at worst.

New hospital, donations of gas money to all kinds of causes.

People spending money on trucks, tractor, polls, additions to their homes.

Downtown stores going out of business.

Huge staging sites with parking lots full of trucks, equipment, temporary buildings.

This iterative and collaborative process informed the development of the Citizen Sense Kit, a collection of different air quality monitors that participants used to monitor air quality in relation to hydraulic fracturing activities. At the same time, it was clear that there was much more to undertaking a citizen-sensing project than distributing monitoring technology within a community. The logbook became a tool kit within a tool kit for learning more about the existing networks of monitoring and action as well as the sedimentations of pollution, politics, and conflict within a distinct area. The logbook framed the how-to as a series of questions, which in turn attended to the communities of inquiry that had formed, and could be in the process of forming, through a citizen-sensing project to study air quality.

From Makerverse to Pluriverse

Making is often discussed as a good or end in and of itself, especially in the sphere of digital technology. Action, getting practical, building, working in a hands-on way: these are proposed as remedies to more sclerotic and inert—indeed, even "academic"—approaches to problems. As Lily Irani has observed in relation to an account of a hackathon, there was a notable "bias for action" in the planning for this event, where hackathon participants "sought to intervene in the operations of the world through 'action' and 'making.'"[10] Such emphasis on making and action can constitute a makerly subject who would undertake activities because they seem productive. Yet, as the pragmatists have discussed and critiqued, action for action's sake is an empty project. Practice is, notwithstanding, the space within which ideas are put to work. It is the very operationalization of ideas—the instrumental aspect of instruments—that the pragmatists stressed was key to how inquiry unfolded and came to have effects, less as the proof of a theory and more as a contingent experience.

In a different but resonant register, Simpson notes that a certain approach to making is part of the integral connection between indigenous knowledge and practices. Making can be "the material basis for experiencing and influencing the world."[11] Her discussion of making and theory is tied to indigenous contexts, and it also produces a philosophy that differently resonates with the pragmatists' approach to practice. Yet Simpson further draws out how the collective undertakings of indigenous politics and governance are also embodied and implemented through distinct forms of making, politics, and governance. She writes,

> Kinetics, the act of doing, isn't just praxis; it also generates and animates theory within Indigenous contexts, and it is the crucial intellectual mode for generating knowledge. Theory and praxis, story and practice, are interdependent, cogenerators of knowledge. Practices are politics. Processes are governance. Doing produces more than knowledge.[12]

The more-than-knowledge that doing produces involves the very relations and networks that make worlds—and these are political inquiries and inhabitations. Just as making is not simply for the sake of making or action, doing is about more than a refinement of theories. Doing unfolds ways of being in and being for worlds.

---

10. Lilly Irani, "Hackathons and the Making of Entrepreneurial Citizenship," *Science, Technology, and Human Values* 40, no. 5 (2015): 807.

11. Simpson, *As We Have Always Done*, 23.

12. Simpson, 20.

Doing can reproduce practices such as settler colonialism. But it can also test, transform, and generate theories in a connected pursuit of the how-to. As forms of doing and action, instruments and instrumentalities are not, in this way, direct lines to certain outcomes but rather constitutive and contingent operations that form worlds.

If we return to the imperative mood, we find that the imperative as procedure here becomes more than instruction. Procedure is always open to revision through ways of living in and making worlds. As the preceding discussion citing Povinelli reminds, making is also about differential ways of being in the world. Making, action, practice, and procedure are about worlds in the plural, the pluralistic universe—or pluriverse—that was the focus of James in his work on radical empiricism.[13] As the "Logbook of Monitoring Practices" example demonstrates, there are multiple ways of monitoring environments and accounting for the effects of pollution through forming tool kits, making sensors, identifying monitoring scenarios, and gathering and analyzing data. How-to can be a way to recognize and support a plurality of modes of inquiry, technical practices, and environmental relations. The makerverse of

13. While James works with the notion of the "pluralistic universe," *pluriverse* is a term that other writers such as Latour have used in relation to James's work. At the same time, writers such as Walter Mignolo take up the pluriverse as a concept but do not cite James as part of the development of this concept, but develop *pluriverse* as a term associated with postcolonial and decolonial theory. Marisol de la Cadena develops yet another reading of the pluriverse through the conservative political writings of Carl Schmitt. This study recognizes these multiple formations of the pluriverse (and somehow it is fitting that this term has a plurality of uses and affiliations) but especially emphasizes James's discussion of the pluralistic universe. See William James, *A Pluralistic Universe* (1909; repr., Lincoln: University of Nebraska Press, 1996); Bruno Latour, *Facing Gaia: Eight Lectures on the New Climatic Regime* (Cambridge: Polity, 2017), 36; Walter Mignolo, "On Pluriversality," *Walter Mignolo* (blog), October 20, 2013, http://waltermignolo.com/on-pluriversality; Marisol de la Cadena, "Indigenous Cosmopolitics in the Andes: Conceptual Reflections beyond 'Politics,'" *Cultural Anthropology* 25, no. 2 (2010): 334–70.

DIY technologies shifts to become the pluriverse of retooled tool kits. The instruments and instrumentalities of sensors are not a unidirectional unfolding of makerly agency but rather networks-in-formation that generate forms of collective causation.[14] Citizen sensing unfolds not just through sensor devices but also in concrete locations and as collective monitoring projects for documenting and addressing environmental pollution. When building a network, you might find, then, that it is helpful to remember that this is an ongoing practice involved in pluralistic fields of relations.

14. For a discussion of collective causation in relation to environmental protest, see Benedikte Zitouni, "Planetary Destruction, Ecofeminists, and Transformative Politics in the Early 1980s," *Interface* 6, no. 2 (2014): 244–70.

How to Test Resistance

WHILE ONCE PRESENTING Citizen Sense research, I was asked by an event participant whether the work was somewhat risky to undertake because it would be perceived as an "activist" project. Indeed, the questioner considered the environmental topic of fracking to be controversial and the perception that the research could be seen to be "helping people" as forgoing the objectivity that is meant to characterize academic research. I have received variants of this question in several other contexts and events, but the basic gist of these inquiries is a worry over the loss of expertise that is seen to be granted by being a distant academic observer and commentator, ideally working on a more neutral research topic.

If ever there were an anecdote well aligned with feminist technoscience, this one surely must seem ready-made to demonstrate the relevance of this body of work. Cue Haraway's "modest witness": the very perception that inquiry involves standing back and letting events take their course, whether in the form of instruments and air pumps or social and political affairs, is a gendered and privileged way of organizing inquiry that allows some people and actions to recede from view to generate universality and objectivity, while others are branded as illegitimate because their presence jams the signal of objectivity.[1] It would be similarly possible to pass through the quantum feminism of Barad to articulate that any observation—even the seemingly most technical and scientific—is an achievement that involves sociopolitical relations.[2] And traveling back to the formation of quantum theory

1. Haraway, Modest_Witness, 23–45.
2. Barad, Meeting the Universe Halfway, 170.

along with its influence on theorists, including Dewey, it would also be possible to say that observing and acting are involved with each other. Observing is acting. Rather than assuming the position of nonengagement to achieve objectivity, Dewey (under the influence of Heisenberg) suggested that new modes of engagement should be deliberately sought to pursue the promise of instrumentalism and philosophy as action.[3] Indeed, as West has pointed out, for Dewey, this was a way in which to ensure "active engagement with the events and affairs of the world" that would contribute to "a worldly philosophy and a more philosophical world."[4]

You might find when taking sensors into the open air, working to build community monitoring networks, and grappling with the environmental problems people seek to address that resistance takes on an electropolitical oscillation. Communities are diversely engaged in struggles, and sensors enter into the fray as part of a process of inquiry and evidence making. Indeed, struggle is a central part of how these projects and practices unfold. Resistance will be encountered not just as a lesson in voltage but also as a response to citizen data, as a query about proper modes of research, and as a question about how or whether governments could be more accountable. Resistance will also be cultivated as part of a process of circumventing established ways of dealing with or overlooking environmental problems, of gathering and presenting evidence consistently and insistently when it is ignored, and of organizing meetings and listening sessions to make citizen observations of environmental problems matter. At the same time, it is important to account for the ways in which expertise differently manifests and how this informs citizen-sensing technologies and practices in the attempt to struggle with accounts of environmental pollution.

The label of activism suggests that the research has forgone its

3. John Dewey, "The Naturalization of Intelligence," in *John Dewey: The Later Works, 1925–1953,* vol. 4, *1929,* 156–77 (1984; repr., Carbondale: Southern Illinois University Press, 2008).

4. West, *American Evasion of Philosophy,* 82.

potential for legitimate inquiry. Yet, on the contrary, because the research is working with and through action and engagement, it is developing new capacities and open-air instrumentalities. Sensors do not simply deliver up transformed political engagements or environmental solutions, despite the marketing promises. Sensors neither singularly empower people nor instantly transform them into activists. Instead, research and practice that are variously situated as collaborative or participatory demonstrate that engagements with environmental problems unfold through differential and complex political struggles. It is by undertaking practice-based and collaborative research that such "findings" become evident. This can also be a way to begin to decolonize research practices and to rework the expert–citizen relations that more colonial modes of research can fix into place.[5]

When working with communities in a participatory way, it is possible to learn about the multiple approaches to addressing environmental pollution, the friction and the discord, the diverse strategies for organizing, and the environmental encroachments that have been held at bay. It is also possible to better understand how collective politics materialize less as a singular pursuit of a goal and more as a working and reworking of instrumentalities: there is work to be done, but the doing of it causes new actions and relations to form. In this way, the political subjects—the instrumental and active citizens—that are constituted through these modes of action are in process. Drawing on the previous discussion of pragmatism, it would be possible to say that action is not the elision of conceptual reflection or development but rather its test and fulfillment. While in the pragmatists' estimation, action is not to be pursued for action's sake, it is also possible to ask what modes of action are under way and what experiences and worlds these would generate.

---

5. For a discussion of how to decolonize methodologies, along with an extensive set of case studies and examples of community research, see Linda Tuhiwai Smith, *Decolonizing Methodologies: Research and Indigenous Peoples,* 2nd ed. (London: Zed Books, 2012).

Activism is one way of parsing action in relation to politics, and yet there is often disagreement about what does and does not count as activism.[6] The how-to is also about modes of action and calls to action. This action can be parsed in many different ways: as action for action's sake or as materialized ways of living. Modes of action are also shifting in relation to present demands. Rather than that old Leninist question "what is to be done?"[7] a more usual question now starts with "how to . . . ?" For some, the question is simply a version of "how to make . . . ?" For others, the question is "how best to live on, considering?"[8] This latter question, raised by Berlant in the context of cruel optimism, is an appropriate one to dwell on at this juncture in this text, because this modality signals most clearly the struggle and resistance that can be embedded within or activated by the how-to.

The search for instructions, the following of procedures, the hopeful pursuit of an effective action or promised outcome: these are ways of looking for direction when potentially floundering on the shores of life. You might find that how-to is a mode of action that often starts in the imperative mood and follows an instrumental trajectory. But how-to is also a vector of transformation. It generates open-air instrumentalities, along with other ways of undertaking research as a collective project. Transformation, nevertheless, encounters resistance and requires struggle. Testing resistance, then, is an important way in which to keep your tool kits well tuned and ready for diverse modes of action, and even activism.

6. Different versions of activism can also emerge in relation to distinct environmental and sociopolitical struggles. For instance, see Anna Lora-Wainwright, *Resigned Activism: Living with Pollution in Rural China* (Cambridge, Mass.: MIT Press, 2017).

7. For a discussion of this question, see Nicholas Gane and Donna Haraway, "When We Have Never Been Human, What Is to Be Done? Interview with Donna Haraway," *Theory, Culture, and Society* 23, no. 7–8 (2006): 135–58.

8. Berlant, *Cruel Optimism*, 3.

How to Retool Tool Kits

I AM FREQUENTLY ASKED whether Citizen Sense research is empowering people and communities through the participatory research undertaken on citizen-sensing technologies. The short answer is, not as directly as that. The medium answer is, we should always query the uncomplicated connection between technology and empowerment. The long answer is, it might be advisable to review this how-to exploration, which seeks to trouble and rework the ways in which tool kits—and empowerment—are constituted. Or, as Stengers has suggested, it might also be possible to consider how to undertake the "empowerment of a situation," which involves "giving a situation that gathers the power to force those who are gathered to think and invent."[1] This how-to guide explains how it might be possible to inhabit yet also to transform the ways in which these technologies operate, through open-air instrumentalisms, and to seek out creative forms of misuse that challenge the assertion that technology made this happen or that Sensor = Outcome. In this way, it might also be possible to reorient the usual and primary attention away from how to make a sensor talk to a platform toward more open-ended engagements with these technologies. Such reorienting and retooling practices challenge the usual configurations of action—and empowerment. They also rework the political relations that are made possible. This is a way in which to retool tool kits.

When using the term *retool,* I am inevitably drawing on work from feminist theory and technoscience to propose how to work

---

1. Isabelle Stengers, "Including Nonhumans in Political Theory: Opening Pandora's Box?," in *Political Matter: Technoscience, Democracy, and Public Life,* ed. Bruce Braun and Sarah Whatmore, 3–34 (Minneapolis: University of Minnesota Press, 2010).

against the grain of dominant technological narratives.[2] Retooling is a practice that is honed through struggle: struggle with and against the standard operating procedures. Retooling is a way to transform and invent technoscientific practices. It asks how tool kits are configured, how they are operationalized, which subjects are drawn into their modes of action, which relations are configured, and which worlds are made and sustained. These are questions of process and mechanism: the how of the how-to so that further engagement and working through of instruction and procedure might find the flex points for transformation.

While this text could have undertaken a survey of citizen-sensing and citizen-science tool kits, I have deliberately opted not to follow the categorical impulse but instead sought to examine the imperative mood and the instrumental mode of action. Rather than pursue a definitional or taxonomic study of practices or tool kits, this text has suggested that tool kits as instruments and instrumentalities give way to inquiry and action that in turn transform the instruments in use. Categories could only ever serve as a provisional way to understand these tool kits and practices.[3] Indeed, the more interesting tool kits incorporate contingency as a crucial part of how they provide resources for organizing action.

Many other tool kits, DIY projects, and community projects have traversed this space of instruction, contingency, action, and alternative engagement. From the Center for Urban Pedagogy's Making Policy Public pamphlet series, which explains and guides publics through complex legal issues like housing and worker's

2. For instance, see Anne Pollock and Banu Subramaniam, "Resisting Power, Retooling Justice: Promises of Feminist Postcolonial Technosciences," *Science, Technology, and Human Values* 41, no. 6 (2016): 951–66, and Ahmed, *Living a Feminist Life.*

3. For a discussion of the limits of categories in relation to citizen science, see Alan Irwin, "Citizen Science and Scientific Citizenship: Same Words, Different Meanings?," in *Science Communication Today: Current Strategies and Means of Action,* ed. Berhand Schiele, Joëlle Le Marec, and Patrick Baranger, 29–38 (Nancy, France: Nancy Université, 2015).

rights,[4] to *A Guidebook for Alternative Nows,* which collects examples of alternative economies and engagement experiments,[5] to Zach Blas's manual for "queer technological strategy"[6] to the *3D Additivist Cookbook* for cooking up alternative inhabitations in troubling times,[7] a wide range of tool kits and guidebooks are experimenting with the form of the instructional and the imperative to work toward more democratic operating conditions.

There are also much different ways of engaging with DIY and technology that can contribute to community projects of addressing environmental pollution and public health. Alondra Nelson has described how the Black Panther Party undertook projects in DIY community health activism, which offered much different ways of mobilizing (medical) technology and political subjects in the interests of social justice.[8] These practices can be ways of transforming technologies and social relations. They make alternative worlds through attending to the political subjects and communities of inquiry involved in open-air instrumentalisms, where experimenting with the conditions and potential of altered technoscientific arrangements can also undo power structures that contribute to health inequalities.

The point of interrogating instrumentality in this way has been to consider how citizen-sensing technologies could be described as instrumental in the limited sense: as merely functional and utensil-like, as Simondon has suggested. Politics as script-

4. Center for Urban Pedagogy, "Making Policy Public," http://welcometocup.org/Projects/MakingPolicyPublic.

5. Amber Hickey, ed., *A Guidebook for Alternative Nows* (Journal of Aesthetics and Protest Press, 2012), https://joaap.org/press/alternativenows_hickey.htm.

6. Zach Blas, *Gay Bombs: User's Manual* (Queer Technologies, 2008), http://www.zachblas.info/wp-content/uploads/2016/03/GB_users-manual_web-version.pdf.

7. Morehshin Allahyari and Daniel Rourke, eds., *The 3D Additivist Cookbook* (Amsterdam: Institute of Network Cultures, 2016), http://additivism.org/cookbook.

8. Alondra Nelson, *Body and Soul.*

ed through these engagements could also be seen to fall into the trap of a more reductive instrumentality. But as this discussion has suggested, there is more to an understanding and practice of instruments and instrumentalities than might initially have been suspected. Instrumentalisms become prospective in the effects they might generate and the relations they might inform. They are formed in the open air, as open-air instrumentalisms. Instrumentalities generate new political inhabitations. The tool kit and the instructional are not necessarily expressive of the starkly functional or extractive form of instrumentality, because instruments develop through engaged and contingent practices. Instrumentalism involves setting in motion, operationalizing, and potentially transforming. Instruments—whether in the form of concepts or sensors—are instrumental to the unfolding, the doing, and the transforming, where other ways of living and other processes are articulated. I have suggested here that *How to Do Things with Sensors* is a project that moves from the imperative mood and reductive instrumentality to one that might generate more contingent open-air instrumentalisms, particularly in relation to citizen engagement with environmental-monitoring technologies. This reworking of citizen-sensing technology and technical relations intends to counter the sinister veneer of Silicon Valley and the smug tyranny of the tech bro, where normativity, exclusion, and reductive technical relations contribute to unjust and undemocratic practices, relations, and worlds.

You might find that once you start to look for instruments, you find them everywhere: much like Haraway's air pump, they are at work in-forming and re-constituting subjects and objects, nature and culture, conditions of fact and knowledge, public and private, legitimacy and illegitimacy, authority and marginality. This guidebook suggests that it would be advisable to approach these instruments through the concept and practice of open-air instrumentalisms, where experimental approaches as well as new technical relations, modes of inquiry, forms of political engagement, environmental commitments, and ways of world-making might materialize. In this sense, *How to Do Things with Sensors* interrogates the diagram of citizen sensing as mode of technological engagement that is meant to lead to specific political effects. Drawing on pragmatism, feminist technoscience, indigenous and critical race theory and practice, this how-to guide develops an approach to sensors where methods, practice, ideas, and theory are co-constituted, embodied, and retooled. How to do things is to ask how to transform things. It is to inquire how to experience and influence worlds. Instruments and instrumentalities do not offer up guaranteed ends but rather unfold operations that are ways of engaging with ideas, technologies, relations, entities, environments, and worlds. Dewey referred to instruments as ideas that were capable of "organizing future observations and experiences" rather than "reporting and registering past experiences." Instrumentalisms, in this sense, are propositional. If there is action to be undertaken, they are in some way focused on making possible interventions and change. This is an approach that focuses on "consequent phenomena" rather than historical facts and is what Dewey would refer to as something "revolutionary in its consequences."[9]

Here's what you might have learned in the process of following these instructions: how-to is not a rule but a proposition. Its imperative mood is one of obligation and even urgency more than command. How-to is an instrumental project, where—following Dewey—meaning arises through contingent operations that make and remake (democratic) worlds. How-to enables modes of inquiry, action, and conduct. How-to is experimental in its searching after ways to address problems. How-to demonstrates how distinct ideas and instrumental actions are tied to different communities of inquiry and possibilities for transforming and retooling action. While the how-to might initially seem to present straightforward instructions pointing toward guaranteed results, the how-to approach should necessarily engage with the pitfalls, de-

9. Dewey, "Development of American Pragmatism," 12.

viations, and antitriumphalism of undertaking citizen-sensing and environmental-monitoring projects in concrete situations. Such a how-to tool kit, then, is productive of open-air instrumentalisms. It works within the genre of the how-to but also seeks to retool this narrative and technical tool kit in order, as Berlant suggests, "to invent new genres for the kinds of speculative work we call theory."[10]

Next, you might find that when working with sensors, you can put these considerations of the how-to to work when asking

1. how to monitor pollution over time;
2. how to teach yourself atmospheric chemistry;
3. how to analyze data;
4. how to construct evidence;
5. how to write a data story;
6. how to ring a regulator;
7. how to influence policy;
8. how to organize a movement;
9. how to remake environmental relations;
10. how to build more just worlds.

10. Berlant, *Cruel Optimism,* 21.

Acknowledgments

This is a necessarily abbreviated list of acknowledgments, given the condensed and developing format of this text; however, I would like to thank the many contributors to the Citizen Sense project, including individual participants, community groups, consultants, collaborators, researchers, and technical advisors. A full list of contributors can be found at citizensense.net. Extra special thanks are due to Helen Pritchard, who helped to make sense of this research and the many splintering—yet ultimately prolific—pathways that it created, and to Kathryn Yusoff, for her generous readings of this text. I received helpful comments from audiences at seminars and lectures at the University of Cambridge, ITU in Copenhagen, the University of Edinburgh, and the University of Exeter. I am grateful for support from Goldsmiths, University of London, where this research began, and from the University of Cambridge, where it now resides. The University of Minnesota Press team has been a delight to work with, and I am appreciative of the support and editorial guidance provided by Danielle Kasprzak and Douglas Armato. This work has benefited from generous support from the European Research Council, which made the Citizen Sense project possible.

Jennifer Gabrys is chair in media, culture, and environment in the Department of Sociology at the University of Cambridge. She is author of *Program Earth: Environmental Sensing Technology and the Making of a Computational Planet* and *Digital Rubbish: A Natural History of Electronics.*